Technikzukünfte, Wissenschaft und Gesellschaft/Futures of Technology, Science and Society

Reihe herausgegeben von
A. Grunwald, Karlsruhe, Deutschland
R. Heil, Karlsruhe, Deutschland
C. Coenen, Karlsruhe, Deutschland

Diese interdisziplinäre Buchreihe ist Technikzukünften in ihren wissenschaftlichen und gesellschaftlichen Kontexten gewidmet. Der Plural „Zukünfte" ist dabei Programm. Denn erstens wird ein breites Spektrum wissenschaftlich-technischer Entwicklungen beleuchtet, und zweitens sind Debatten zu Technowissenschaften wie u.a. den Bio-, Informations-, Nano- und Neurotechnologien oder der Robotik durch eine Vielzahl von Perspektiven und Interessen bestimmt. Diese Zukünfte beeinflussen einerseits den Verlauf des Fortschritts, seine Ergebnisse und Folgen, z.b. durch Ausgestaltung der wissenschaftlichen Agenda. Andererseits sind wissenschaftlich-technische Neuerungen Anlass, neue Zukünfte mit anderen gesellschaftlichen Implikationen auszudenken. Diese Wechselseitigkeit reflektierend, befasst sich die Reihe vorrangig mit der sozialen und kulturellen Prägung von Naturwissenschaft und Technik, der verantwortlichen Gestaltung ihrer Ergebnisse in der Gesellschaft sowie mit den Auswirkungen auf unsere Bilder vom Menschen.

This interdisciplinary series of books is devoted to technology futures in their scientific and societal contexts. The use of the plural "futures" is by no means accidental: firstly, light is to be shed on a broad spectrum of developments in science and technology; secondly, debates on technoscientific fields such as biotechnology, information technology, nanotechnology, neurotechnology and robotics are influenced by a multitude of viewpoints and interests. On the one hand, these futures have an impact on the way advances are made, as well as on their results and consequences, for example by shaping the scientific agenda. On the other hand, scientific and technological innovations offer an opportunity to conceive of new futures with different implications for society. Reflecting this reciprocity, the series concentrates primarily on the way in which science and technology are influenced social and culturally, on how their results can be shaped in a responsible manner in society, and on the way they affect our images of humankind.

Weitere Bände in der Reihe http://www.springer.com/series/13596

Arie Rip

Futures of Science and Technology in Society

 Springer VS

Arie Rip
University of Twente
Enschede, The Netherlands

ISSN 2524-3764 ISSN 2524-3772 (electronic)
Technikzukünfte, Wissenschaft und Gesellschaft/Futures of Technology, Science and Society
ISBN 978-3-658-21753-2 ISBN 978-3-658-21754-9 (eBook)
https://doi.org/10.1007/978-3-658-21754-9

Library of Congress Control Number: 2018946684

Springer VS

Verantwortlich im Verlag: Frank Schindler

Printed on acid-free paper

This Springer VS imprint is published by the registered company Springer Fachmedien Wiesbaden
GmbH part of Springer Nature
The registered company address is: Abraham-Lincoln-Str. 46, 65189 Wiesbaden, Germany

Preface

Christopher Coenen and Armin Grunwald

Arie Rip is one of the most internationally renowned scholars working on issues of science, technology and society. His groundbreaking work has been highly influential in many areas of inquiry, and has stimulated a wide range of research in science and technology studies (STS), technology assessment (TA) and adjacent fields. It has inspired a large number of PhD theses, enriched numerous conferences and workshops, and fueled many discussions and debates.

He has also been a leading voice for decades when it comes to the topic of our book series, "Futures of Technology, Science and Society", shaping the discussions about relevant fields of new and emerging science and technology at the intersections of STS, TA and other areas of study, as well as policy advice at both the European and the international levels. We are therefore delighted that a collection of important essays by Rip, which give insights into the evolution of his thought in recent years, is now being published in this series. We believe it will be highly beneficial for further research, education and public communication on science, technology and societal futures.

With two exceptions – the updated introduction and an important paper about responsible research and innovation (RRI) –, the essays included in this volume appeared previously in a booklet handed out to the participants at the symposium "Future of Science and Technology in Society", which was organized by the Department of Science, Technology and Policy Studies (STePS) and the Institute of Innovation and Governance Studies (IGS) at the University of Twente, and held on 16-17 June 2011. This event marked the passage of five years since Arie Rip formally retired, and the title of the booklet was "Futures of Science and Technology in Society", using the same plural form as we chose for our series. We wish to thank Stefan Kuhlmann, Chair of STePS (and co-organizer of the 2011 Symposium), and the IGS for their permission to use the booklet for this publication and make these works more widely accessible.

We are confident the present volume will not only be warmly welcomed by scholars and scientists interested in Arie Rip's thinking, but also fits in perfectly with the concept, topic and spirit of the "Futures of Technology, Science and Society" series, and will help us to develop it further. In his introduction to this collection, Arie Rip cites the famous remark made in 1959 by C.P. Snow that scientists have the future in their bones. In the meantime – and Rip has made a crucial contribution to this development –, a strongly interdisciplinary culture of anticipation has emerged in discourse on science, technology and society. In the introduction, Rip explains how anticipatory thought in STS and TA has increasingly contributed to the governance of new and emerging science and technology, promoting methods that allow for higher degrees of reflexivity. Such methods may be based on prognostic work, but above all help improve governance processes, supporting procedural innovation and enabling comprehensive approaches. With the present volume, the communities that are seeking to foster this culture of anticipation now not only have a resource at their disposal they can regularly use in pertinent future work and refer to in deliberations, but at the same time a source of inspiration for continuous reflection on their own practices and the future-oriented governance of science and technology in general.

Contents

Introduction

Social scientists (as well as philosophers and intellectuals for that matter) are tempted to offer a distantiated view, from above – and thus from nowhere –, and do so by inclination and/or to avoid biases. The other extreme, to identify with actors, as in versions of action research and so-called normative approaches, is not a real alternative. My approach has been to see the analyst (myself) as embedded in the same world as actors, but moving about in it in a different way, circulating across sites, and thus seeing different things, or at least, seeing things differently. This then is also an entrance point to think and write about, and sometimes engage in, how things might go differently.

The future is everywhere. As C.P. Snow noted in his 1959 lecture about the 'Two Cultures' (Snow 1961), scientists have the future in their bones. New technologies live on promises. And we tend to tell ourselves forward. Anticipation is integral to actions and interactions, and can be made explicit – from the informal scenarios embedded in the stories in which we position ourselves and others, to justifications of science and technology policy and the forward-looking versions of technology assessment. Such anticipations can be done reflexively, by seeing them as embedded in ongoing developments, and thus part of de facto governance, rather than as forecasting or foresight exercises that can be optimized as such. This is the broader background for my selection of articles about the futures of science and technology in society.

In my intellectual and scholarly work, three overlapping lines of analysis and diagnosis can be distinguished. In the first line, about science dynamics, I developed an overall approach. I look at knowledge production and scientific institutions as an evolving ecology/landscape in which the analyst is situated as well. To understand science and think about its future, I went back to an ecology/landscape where science as we know it now was not visible: the 16th and early 17th century in Europe, and then moved up to the present and articulated some of its "endogenous" futures. (Chapters 1 and 2.) This approach actually integrates lots

© Springer Fachmedien Wiesbaden GmbH, part of Springer Nature 2018   1
A. Rip, *Futures of Science and Technology in Society*, Technikzukünfte,
Wissenschaft und Gesellschaft / Futures of Technology, Science and
Society, https://doi.org/10.1007/978-3-658-21754-9_1

of work in STS and in science policy studies, as I showed in an unpublished paper available on my website (Rip 2005).

For technology and technoscience, and their situated dynamics (the second line in my work), an ecological approach is also possible but much more complex, because compared with science there is much less institutional autonomy. In my work I have focused on the dynamics of technological development and its embedding in society. I developed a multi-level, co-evolutionary perspective which has informed my analysis, and that of others (up to the so-called multi-level approach to sociotechnical transitions, particularly towards sustainability). Co-evolution of technology and society is the lens that allowed me to articulate new/better conceptualizations of technology development and its governance. (Chapters 3 and 4.)

In addition, I developed ways to increase reflexivity of the co-evolution of science, technology and society, in particular Constructive Technology Assessment. (Chapter 5.) My work over the last decade (together with PhD students) on Technology Assessment and societal aspects of nanotechnology is an example. It is also an example of how one can be partially embedded in the world of actors, in this case the R&D consortium NanoNed in the Netherlands and European Networks of Excellence. And it reminded me of how I started out in STS in the 1970s: teaching and researching chemistry and society in a chemistry department (in the University of Leiden). At the time, one of the arguments for having such teaching, and having it in a science department, was the reference to the Dutch Higher Education Law, enjoining universities to pay attention to the advancement of societal responsibility (of students). By now, a broader version of this argument has become visible in the call for Responsible Research and Innovation, for example in the European Commission's Framework Program Horizon 2020. While it has some features of a fashionable policy concern, it should be seen as a social innovation with a still uncertain future. I have actually returned to a theme I developed my 1981 PhD thesis on Societal Responsibility of Chemists: evolving divisions of moral labour. I have shown how the notion of Responsible Research and Innovation creates openings for reflecting on present divisions of moral labour and attempts to modify them (Chapter 6). And I have been involved, in an advisory role, in the European Commission's attempts to put Responsible Research and Innovation on the map.

I am a reflexive person, as was clear already in the opening paragraph of this Introduction. This has led to a third line in my intellectual and scholarly work, addressing basic sociological and ontological issues, seriously as well as playfully, and showing my preference for ironic engagement. Most of it is unpublished, but I offer a published piece on material narratives (Chapter 7) which has been appreciated by philosophers of technology. And an unpublished text that is dear to me, as an "exhibit" (Chapter 8). When applying for the chair of Philosophy of Science and

Technology at the University of Twente in Spring 1987, I gave a lecture in which I outlined a perspective on social order and social ordering. I revisited this lecture in a colloquium I gave just before my formal retirement in 2006; it turned out I did not need to change its thrust. Did I not make much progress over the intervening years? I prefer to see the 1987 lecture as an attempt to capture important but neglected aspects of social ordering that I did not get around to develop much further. Eventually, I wrote up an English version of the 1987 lecture, with 2006 additions, which now constitutes Chapter 8. The thinking behind it has informed some of my work, including a recent paper on the illusion of risk control.

The perspective is informed by STS, in particular my version of Actor-Network Theory where "entanglement" is the basic dynamic. (cf. Rip 2010). Actor-Network Theory can be pushed as a polemics with mainstream social science, as Bruno Latour tends to do. I prefer to position it as an additional layer of understanding dynamics of social ordering.

Looking backward, I realize how much I profited, over the years, from discussions and collaborations with colleagues/friends. Such enjoyable intellectual interaction often emerged already in PhD student – superviser relationships (and sometimes with adopted PhD students). I am grateful that this was possible, and I hope to continue and enjoy our interactions.

Looking forward, I will continue to develop the three lines of analysis and diagnosis I outlined as characterizing my work. There is still so much to say and to do. This book offers some achievements, and in doing so, indicates the intellectual platform from which I will continue to work.

References

Rip, Arie, Haven't we got all the theory we need? an informal paper prepared for the workshop Middle Range Theories in Science and Technology Studies, Amsterdam, 27-29 April 2005

Rip, Arie, Processes of Entanglement, in Madeleine Akrich, Yannick Barthe, Fabian Muniesa et Philippe Mustar (réd.), Débordements. Mélanges offerts à Michel Callon. Paris: Transvalor - Presses des Mines, 2010. pp. 381-392

Snow. C.P. The Two Cultures and the Scientific Revolution. The Rede Lecture 1959 .The Syndics of the Cambridge University Press , 1961

Chapter 1
Protected Spaces of Science: Their Emergence and Further Evolution in a Changing World*

Introduction

Most often, discussions of ongoing changes in science in society are framed, by actors as well as analysts, in terms of science-as-we-know-it. In fact, the reference is often to science as-we-knew-it, to a Golden Age when things were better. Indicative is how US President Obama's phrase, in his inaugural address in January 2009, about "restoring science to its rightful place", was taken up by scientific establishments. The phrase was meant to contrast with the Bush Administration's politicization of science,[1] but spokespersons for science picked it up and interpreted it as "more money, more freedom for science". This shows the deeply engrained "entitlement" attitude of scientists, where the structural dependence of science on sponsors is backgrounded, and turned into a "right."

The origin of this "entitlement" attitude can be traced back to the 1870s, with the various "endowment of science" movements in the UK, France and Germany. In other words, it is historically contingent and its force derives from the eventual institutionalization of certain sponsorship constellations, not from characteristics of science as such. Having seen this, one starts to wonder whether there can be something like "science as such", somehow given, independent of history. There are enduring achievements, but science, as we know it now, is also the convergence (i.e. inclusion and exclusion) over time of different activities, their institutionalization at particular times and places, and their further co-evolution.

---

*   Source: Chapter in Martin Carrier & Alfred Nordmann (eds.), *Science in the Context of Application: Methodological Change, Conceptual Transformation, Cultural Reorientation,* Dordrecht: Springer, 2011, pp. 197-220.

1   Cf. the March 9, 2009, Memorandum to Heads of Agencies, on scientific integrity. http://www.whitehouse.gov/the_press_office/Memorandum-for-the-Heads-of-Executive-Departments-and-Agencies-3-9-09/ accessed 17 March 2009.

© Springer Fachmedien Wiesbaden GmbH, part of Springer Nature 2018
A. Rip, *Futures of Science and Technology in Society*, Technikzukünfte,
Wissenschaft und Gesellschaft / Futures of Technology, Science and
Society, https://doi.org/10.1007/978-3-658-21754-9_2

This is not a message of relativism. What has (co-)evolved over time has value, and there are important issues at stake in the present discussions. What I want to problematize is the simplistic reification of science as something given, somehow, which can then also be referred to as a standard, as what is "proper" science. Of course, achievements must be recognized and cherished, and when threatened, defended and hopefully "restored." But one has to consider possible further evolutions, and their value, also if this does not conform to what is now considered "proper" by scientific establishments. Standards for evaluation cannot be specified beforehand, but co-evolve with practices and institutionalizations. Still, there is a continuing thread, the goal and practices of robust knowledge production (in context). I will come back to this in the next section, and build on it to offer my diagnosis.

To do so, I have to clear away pre-conceptions about science and its dynamics. Science is not just a way (perhaps the main way) of producing robust knowledge, it is also part of a master narrative of progress, and has become an icon of modernity. And it has become linked to nation states, which sponsor scientific research, and shape its "rightful place". This is what the Bush Administration did (even if one may not be happy with it) and what the present Obama Administration does. What is done at the laboratory bench (and increasingly, in the computer), is not independent of these larger developments, even if the scientists, in their protected spaces in the lab, do not feel the impacts directly. I will develop this point by showing the importance of protected spaces, not just at the micro-level of the laboratory, but also at the macro-level of a "rightful place" for science in society, and at the meso-level of scientific communities and institutions of the science system.

The epistemic and institutional aspects are entangled, at the micro-, meso- and macro-levels. This is already visible in how Kuhn (1970), in his Postscript, emphasizes that the (epistemic) paradigm and the relevant scientific community are two sides of the same coin. A further point was introduced by Campbell (1979): in such scientific communities there are "tribal norms" (like struggle for visibility) which may not have an immediate epistemic value, but support the life of the community, and are thus important for knowledge production, and shape it. One can see the epistemic and the institutional as two different dynamics which impinge on each other, and may, or may not, support each other. In fact, they are integral to each other.

This perspective implies a criticism of much of philosophy of science: while the importance of social and institutional aspects is increasingly recognized, it is taken as a context, and thus external to the core, epistemic business of science, rather than an integral part of epistemic practices. The sociology of science should be criticized as well, however, for its neglect, or at least black-boxing, of the epistemic business of science. There appears to be a division of intellectual labour here. Philosophy of science looks at what is happening within the protected places at the micro-level and at the meso-level of disciplines, and forgets to ask about the nature and effects of the protection. The Mertonian sociology of science (Merton 1973) stays outside, while laboratory studies immerse themselves within it and forget about the outside (as in Latour and Woolgar 1979, where the specifics of biomedical science in the USA in the 1970s are not discussed).

This is a bit of a caricature, because there is lots of interesting work done that transcends these strong reductions of complexity (and I can build on such work for my analysis). But the caricature does indicate that I have to battle on two fronts: integrate the institutional in the epistemic focus of philosophers, and integrate the epistemic in the institutional focus of sociologists.

Many of the current diagnoses of changes in science and its interactions with society focus on institutional aspects, as in the idea of university, government and industry overlapping and co-evolving as in a Triple Helix (Etzkowitz and Leydesdorff 2000).[2] Closer to my call for an integrated socio-epistemic approach is the diagnosis of wide-ranging changes in modes of knowledge production put forward by Gibbons et al. (1994) and Nowotny et al. (2001).

Gibbons et al. (1994) contrast an earlier Mode 1 (university-based and disciplinary oriented) with a presently emerging Mode 2, which is transdisciplinary, fluid, has a variety of sites of knowledge production including "discovery in the context of application" (e.g. in industry) and new forms of quality control. The separate features they describe are clearly visible, but one might want to question their overall thesis that these add up to a new mode of knowledge production, comparable in its internal and external alignments and eventual stabilization to Mode 1 (Rip 2000a).

More important for my analysis and eventual diagnosis is the recognition that their Mode 1 is historically located. Its building blocks emerged during the 19th century, and these became aligned, and locked-in after 1870 (as I will discuss later). However, there was science, or at least robust knowledge production, before the 19th century. If one wants to specify encompassing modes of knowledge production, one could say there must have been a Mode 0 of knowledge production. There might not have been a specific mode of knowledge production, though, rather overlapping varieties of knowledge production, as in the "melting pot" of the Renaissance in Europe.

I will address these issues in the next sections in terms of identifiable contextual transformations which are followed by stretches of more or less incremental development. A basic question, important for the diagnosis of our present situation, is visible already. How could a Mode 1 emerge at all and get a hold on the variety of

---

2   See Hessels and Van Lente (2008) for an overview and for a discussion of the reception of the Gibbons et al. (1994) claim about a new mode of knowledge production.

knowledge production and institutions? The key "mechanism" I propose is a lock-in of dynamics at three levels: ongoing search practices and knowledge production "on location", more cosmopolitan interactions of scientists (and practitioners more generally) and the institutional infrastructures to do so, and legitimation of science and its role in society. Such a lock-in creates nested protected spaces for doing science, and in a particular way – in the case of Mode 1, the combination of relative autonomy and disciplinary authority --, at the price of accepting the constraints that go with such protection. One such constraint is the hold disciplines have obtained on the production of scientific knowledge. Another constraint derives from the norms and values dominant in the regime of Science, The Endless Frontier, visible already from the late 19th century onwards, but coming into its own after the second world war (Bush 1945). The entitlement attitude identified in the opening paragraph is part of this regime.

Clearly, we need a long-term perspective to offer an adequate diagnosis of ongoing changes in science in society.

Long-Term Dynamics of Institutionalized Knowledge Production

In a long-term view, one must be careful in speaking of 'science' because it is only from the early 19th century onward that an easy reference to science is possible. True, the word 'science' was used before, but it was only one of a range of terms, including "natural philosophy".[3] Still, one needs some guideline as to what to include in the analysis of developments. To indicate continuities, or at least lineage, one might still speak of 'scientific' knowledge production, but using quotes as a reminder that the term science refers to eventual institutionalizations, and may not have been used at the time.

To cover the variety of modes of knowledge production, a broad description is necessary. I will just state the key elements, but they can be argued in more detail (cf. Rip 2002b). Given the dominance of science as presently institutionalized, some

---

3   Indicative is that the word 'scientist' was coined by Whewell in 1833 "to designate collectively those who studied material nature." Morrell and Thackray (1981, 20) locate this as a response to Coleridge's challenge to the 1833 Cambridge Meeting of the British Association, that the members should not call themselves philosophers. Ross (1962) gives the story of the word.

of my formulations could be read as polemics with the strong claims about science as the exclusive road to valid knowledge.

Knowledge that claims some validity, scientific or otherwise, is a precarious outcome of efforts to make knowledge applicable at other places and other times – so that one can learn from one place and time to another, and act on that knowledge with some confidence. When knowledge production becomes professionalized, such "acting" includes its use in further knowledge production.

The transformation of local experiences to findings with a cosmopolitan status is an essential ingredient of the 'scientific' mode of knowledge production: it is the (precarious) basis of scientific claims of universal validity. Such transformations are not limited to the specific mode of knowledge production of modern western science, however. Professional knowledges are one example, and craft knowledge and folk knowledge can also work towards cosmopolitan status.

The claim of the applicability elsewhere and elsewhen of the knowledge produced raises two general questions. One is how robust results are produced on location. To get nature to work for us, and on our terms, whether in scientific experiments, industrial production, or agricultural and health practices, we have to shape it, and use whatever comes to hand. Already in the creation of a laboratory and in the set-up of experiments, local and craft knowledge are important, and thus form an integral, albeit neglected, part of scientific knowledge production.

The other is how cosmopolitan knowledge can be translated back to concrete situations (and how to operationalize the notion of validity). The movement for evidence-based medicine offers an interesting case, showing the ambivalences, because it transcended and improved upon local, experience-based knowledge, but has now "overshot the mark" and "excludes too much of the knowledge and practice that can be harvested from experience (..) reflected upon"(Berwick 2005).

Institutionalization of knowledge production implies the emergence of tried and trusted ways of producing knowledge that can claim to be valid. A phrase like "disciplined enquiry" captures this (Kogan 2005, 19), but also indicates the ambivalence involved, when institutionalized disciplines start to discipline ongoing practices of knowledge production. Already within science as-we-know-it there is variety, in particular between more experimental approaches and more "natural history" approaches. The unity of science is primarily institutional.

This outline of a philosophy and sociology of knowledge that claims validity is the backbone of my analysis of developments in knowledge production and its institutionalizations. Taking a bird's eye view, one can identify major changes as well as periods of relative continuity. Mendelsohn's diagnosis of three main transformations remains relevant (Mendelsohn 1969), and I will follow his lead but speak of contextual transformations to do justice to the entanglement of the epistemic and the institutional. Later research has corroborated the diagnosis of a "positive" transformation in the second half of 17th century (see especially Van den Daele 1977a and 1977b) and a "professional" transformation in the course of the 19th century (which leads to the lock-in of Mode 1). In the late 20th century, the earlier regime opens up. New closures that emerge might add up to a third transformation, but it is unclear what it might consist of.[4]

The diagram below (Figure 1) offers a (selective) overview of long-term socio-epistemic developments. In the diagram, I use the notion of a "social contract" between science and society to identify a key element in the transformations and their stabilization, even if it is not a formal contract, and the partners of the contract are ill-defined.[5]

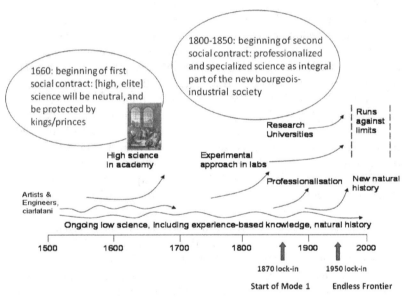

Fig. 1   Long-term developments of 'scientific' knowledge production

4   At the time (Rip 1988), I spoke of a political transformation, but that was programmatic. By now, there are some indications, if one takes "political" to mean the increased and explicit interaction of society with science.

5   The notion of a social contract between science and society has been used before, particularly in the USA, and offers a way to diagnose what is happening now as the breakdown of an earlier social contract, and then identify elements of a new social contract (Guston and Kenniston 1994).

In the next three sections I will zoom in on some parts of the socio-epistemic history which are relevant to my search for a diagnosis that includes a long-term perspective. Here, I note five features of the overall history which are always relevant, even when not foregrounded.

First, the ever-present messiness and heterogeneity (socially and epistemically), which is more visible in natural history than in laboratory-experimental approaches.

Second, the movement from local to cosmopolitan, and back again, where social/institutional and epistemic features are two sides of the same coin. This was visible already in my broad description, above, and can be developed further[6]

Third, the key role of sponsors in enabling knowledge production and shaping its institutionalization. Different forms of patronage occur over the centuries, and include present science policy and university-industry interactions.

Fourth, the emergence of partial and sometimes hegemonic regimes, where macro and micro are aligned. Often, through intermediate (or meso) level organizarions or institutions, from the scientific societies in the 17th and 18th century to the research funding organizations of the 20th century.

Fifth, the creation (and increasing importance) of protected spaces. At the macro-level, protection by kings (as in the 1660s in Britain and France) and later by nation states. At the micro-level, laboratories and controlled experiments. At the meso-level, intermediary structures like the funding-agency world after the second world war.

Nested protected spaces are the distinguishing characteristic of knowledge production in science-as-we-know-it. Protected spaces have material, socio-cultural, and institutional aspects. This is clear in the notion of a laboratory as a place where experiments can be conducted under restricted conditions: these conditions include the disciplining of its inhabitants and the exclusion of unwanted visitors. Field sciences have more difficulties in creating the desired protection, but attempt to create their boundaries as well, especially when aspiring to be part of high science.

The effect of protected spaces is the reduction of interference <u>and</u> of variety. In other words, productivity of scientific knowledge production is based on exclusion. This holds for laboratories (and their equivalents) and for disciplinary scientific communities which guard their status by excluding those who are not qualified. And

---

6   As I have shown (Rip 1997), going from the local to the cosmopolitan is an epistemic and a social (institutional/political) movement. It involves circulation (among localities, possibly guided by cosmopolitan rules), aggregation (forums, intermediary actors), and an infrastructure for circulation & aggregation. Generally valid knowledge can only be achieved when there is a functioning cosmopolitan level. The nature of the knowledge produced will then be shaped by the affordances present at the cosmopolitan level (cf. Campbell's (1979) point about tribal norms).

for professionalized, authoritative science (since late 19th century), which excludes other loci and modes of knowledge production as non-scientific.

Thus, there is an essential tension: the productivity of scientific knowledge production is based on exclusion, and this may reduce unruliness and innovation. In an earlier attempt to address and diagnose socio-epistemic changes, I noted:

> (..) the recurrent and unavoidable dilemma between – on the one hand – the need for some order, and the reduction of variety that goes with it to be productive in what one does (here, search for knowledge), and – on the other hand -- the need to go against that same order to innovate, or just to respond to changing circumstances. For science, and its institutionalized interest in producing novelties (up to priority races and conflicts), the dilemma is an essential tension. [As Kuhn (1977) phrased it and Polanyi (1963) experienced it.] (Rip 2002b, 101)[7]

The dilemma cannot be resolved, but it is made tractable in practice. Protected spaces which enable as well as constrain make it tractable. Their existence has become a functional requirement for doing science, but the specific ways in which these spaces enable and constrain can have limitations, or may even be counter-productive. Also, there are pressures from without as well as from within on existing protected spaces: to change, to become porous, perhaps to be abolished. A diagnosis should be based on an analysis of long-term developments, so as to understand the nature and functionality of the protected spaces. This part of the diagnosis then leads to further questions: are present protected spaces opening up? are new kinds of protected spaces emerging? I will address these further questions (albeit selectively) in the last sections of this chapter.

The Melting Pot of the Renaissance and Partial Closures

For a birthplace of Western science as-we-know-it, 14th-16th century Renaissance Europe looked messy, unruly, and without clear boundaries between various knowledges. There were the (medieval) universities. There were travelling humanists, artists and engineers. There were also almanac makers, astrologers, mountebanks and *ciarlatani* performing tricks at the fairs. Princes and wealthy persons were sought as

---

7   I also offered a diagnosis: "Science, in its interest in searching for knowledge and trying to make its products robust, can be contrasted with science as an authority, which often relies on traditional ways of knowledge production and disciplinary controls of quality. If authority as such, disciplinary or otherwise, rules, science becomes its own worst enemy."

sponsors. The scholarly and craftwork to be done was defined in terms of the wishes and aspirations of sponsors, as well as for the market place.[8]

The variety of knowledge production visible in the Renaissance became partially contained. Micro-protected spaces for experiments were introduced by Boyle and others: somewhat controlled conditions, and oriented towards demonstration. This was combined with macro-protected spaces (privileged by a King) where 'deviant' approaches were excluded. At the same time, natural-history approaches to knowledge production continued -- valued by sponsors because it allowed exploitation of what is 'out there'.

The so-called scientific revolution of the 17th century replaced unruliness with proper procedure (in scientific academies) and started to create boundaries between mechanical philosophy and the crafts (Van den Daele 1977a, 1977b). While this was just one part of the developments, the distinction between "high science" and "low science" (as I have called it, referring to a similar distinction between Anglican high church and low church) would continue, up to the eventual dominance of physics in the pecking-order of disciplines. Whether one considers this development as an achievement or as de-humanisation (Toulmin 1990), the rationalistic mode of knowledge production which eventually emerged had grown out of the fertile soil of the Renaissance. The richness, variety and openness of knowledge production at the time were important for the scientific revolution. And I add, it remained, and remains, important as a backdrop to high science, and as a source of renewal.

Within this overall shift, sponsors in interaction with scholars and artists played an important role, and this is also how a key institution of modern science, peer review, emerged. In Renaissance Europe, immediate and bilateral patron-client relationships developed into triangular relationships, in which the patron needed advice about his sponsorship of a painting, a sculpture, or an engineering work, from a knowledgeable third party – in particular, humanist and other Renaissance scholars, who might on other occasions profit from patronage themselves. This circulation enabled the emergence of a community of what we now call "peers", and the practice of "peer review" – which remains, essentially, advice to a sponsor, i.e. a journal editor/publisher or a research funding agency (Rip 1985).

---

8   One intriguing variety of knowledge production was through so-called 'professors of secrets'. They collected recipes from different crafts and some of their own experience, and sold them on the fairs or to sponsors. The ambivalence in their position is curiously similar to that of biotechnologists and other scientists in commercially important areas. They had to advertise themselves and their knowledge in order to create some visibility. However, at the same time they had to keep their secrets in order to maintain a competitive advantage over other such 'professors' operating on the same market or for the same sponsors (Eamon 1985).

In the case of Galileo at the court of the Medici in Florence, this is visible, and further patterns can be seen emerging that eventually became a fact of 'scientific' life. As Biagioli (1993) showed in detail, Galileo was first of all a courtier who offered his work to his patron, and looked carefully after his "local net", but he was also active in building a "cosmopolitan net" with his competing colleagues at other courts – the competition focused on who could offer the more interesting things to their respective patrons --, and distancing himself from other, low-brow clients of his patron.

Cosmopolitan interactions, while deriving from, or at least coupled to, local contexts and interests, stimulated the emergence of virtual communities, linked through circulating texts and their contents. The influence of patronage games continued in a more global way, as when institutional etiquette was enforced. The need to appear courteous pushed the struggles among practitioners below the surface that was presented to the outside (cf. Shapin 1994). To coin a phrase: Scientists are tradesmen rather than gentlemen, but need to behave, and seen to behave, nicely to keep up legitimation.[9]

From the late 17th century onwards, the emergence of scholarly journals in the Republic of Letters helped to support "cosmopolitan nets". The scientific societies of the eighteenth century could publish reports of research, and might channel support from patrons to their members. The Enlightenment movement (in its various instantiations in different countries) allowed for overall legitimation of scientific knowledge production, independent of the support by specific patrons. At the same time, specific practices, e.g. of mining and metallurgy, or medical preparations, or meteorological data collection, were developing general insights, and thus added a cosmopolitan level as well. This could link up with general theorizing, as in the case of chemistry, and thus create proto-disciplinary communities (Hufbauer 1982).

Professionalisation of Science in Bourgeois-Industrial Society

While the history of the emergence of disciplines and specialties starts in the late 18th century they become a serious business with professionalization of science and the revitalisation of higher education in the second half of the 19th century. By the late 19th century, disciplines were becoming dominant institutional categories,

---

9   This difference between public presentation of science and actual interactions inside the world of science continues, cf. Gilbert and Mulkay (1984) on contingent and rational repertoires.

sedimented and codified in university departments and library categories. This is the institutional infrastructure for recognized specialties to emerge, with their own paradigm, cognitive style, and ideals of explanation.

Part of the work in research practices then becomes to transform the local production of knowledge items into more cosmopolitan knowledge claims – as in a scientific paper. Such claims are addressed to non-local audiences as constituted by a research area. These audiences and areas can be hybrid, as was (and is) the case in many sub-areas of chemistry (Rip 1997). Research areas, specialties and disciplines offer spaces for cosmopolitan scientific work-- a protected space at the meso-level.

Scientific work became sufficiently independent to relate to, and profit from, distributed sponsorship: from scholarly societies, various patrons, the state (in particular in France and in the German states) and professional practices (as in the UK). The 1870s mark a further change. Spokespersons for "science" felt sufficiently secure to claim that "science" should be "endowed" by the nation state (MacLeod 1972). The state responded and became a general sponsor. In parallel, universities started taking up research and scholarship in earnest.

The increased role of the nation state strengthened the idea of a national community of scientists, located primarily at universities. While there had been self-styled spokespersons for science before, there now emerged a scientific establishment with institutionalized channels for lobbying and advice. This partial lock-in became complete when government funding agencies for science took off after the second world war: the agencies were captured by the national scientific communities, legitimated by the ideology of "Science, The Endless Frontier", which could now dispense resources (Rip 1994). In a phrase: scientists divided the spoils (while voicing concerns about insufficient funding). Funding agencies became the bastion of disciplines, although with occasional, and now increasing, guilt feelings about multi- and interdisciplinary work, and attempts to respond to new developments. The authority of disciplines thus derives from the combination of their ordering of knowledge production, and their role as sponsoring categories in national research systems.

Sponsors and Spaces

This history of the emergence of Mode 1 shows how sponsorship of science is an integral element. A closer look at the variety of sponsorship relations actually indicates that there was always more to science than the regime of Mode 1. This allows me to introduce a further aspect of the dynamics of the development of science, which became important in the late 20th century.

Since the late 19th century, local and state governments and industrial firms have used research and researchers for particular services, employing them or contracting them. An element of sponsorship was added because of the expectation of general value of the findings (so no detailed specifications of the work) and because the researchers were allowed to further their own reputation and career. This worked out differently in different scientific fields. In chemistry, from the late 19th century onward, a productive practice developed of interactions with industry and other sponsors, including a workable etiquette, particularly since the interbellum.[10] In fact, this allowed chemists to accommodate the new challenge of biotechnology in the 1980s and 1990s.[11]

The big charitable foundations, first established in the early twentieth century, are the nearest equivalent to the earlier patrons of science who could, and would, act according to their own discretion. The Rockefeller Foundation, based in the USA, had a generalized interest in natural and social science, linked to its concern about the future of urban-industrial society. It has stimulated new developments in biology (including work that paved the way for molecular biology), anthropology and social science from the 1930s until at least the 1960s. Being funded by the Rockefeller Foundation added to the reputation of the researcher and the research institution.

In addition to such concrete sponsors, one can see the emergence of abstract sponsors, starting with the idea (or ideograph, cf. Rip 1997) of SCIENCE as progress through the advancement of knowledge. Reference to this abstract sponsor supports concrete resource mobilisation efforts, especially with the state and with science funding agencies, and is thus an indirect source of resources. The nation state, a concrete sponsor from the 1870s onwards, also became an abstract sponsor

---

10 The wishes of customers and sponsors were internalized in the field, that is, need not be present as such to have an influence. The functionalities the sponsors were interested in would be realized through the heuristics that made up the paradigm or the regime (Slack 1972; Van den Belt and Rip 1987). This way of formulating the point resembles the finalisation and functionalization thesis of the Starnberg group (Schäfer 1983), but does not depend on their overall (and physicalist) diagnosis of the development of Western science.

11 Biologists, on the other hand, had no such history of interaction with industry (their practical relations with sponsors were in medical and agricultural sectors), so the advent of biotechnology created transitional problems, with conflicting etiquettes and complaints of naiveté (Rip and Van Steijn 1985).

when scientists started to refer to their duty to the NATION, which would shape directions of their research (in return for support).[12]

The emergence of a further ideograph, INDUSTRY, is very visible in chemistry. In addition to Bayer, Hoechst, ICI or Dupont contracting for specific types of research, it was toward the chemical sector in general (and also the medical and pharmaceutical sectors) that chemical research and researchers would be oriented, explicitly or implicitly. Since reference to the importance of industry helped to mobilize resources, the ideograph INDUSTRY became an abstract sponsor. Reference to INDUSTRY was increasingly important for science in the late 20th century. Spokespersons for industry (that is, INDUSTRY) were expected to sit in committees, and chairmen of science funding agencies are often required to have some experience in industry, or at least experience in the private sector.[13]

There are other such combinations of concrete and abstract sponsors, the MILITARY being a prime example in the post World War II situation – even if the link of science to the MILITARY is now also contested. The ideograph SUSTAINABILITY has become powerful in recent years. NGOs (non-governmental organisations) ranging from Greenpeace to the International Council of Scientific Unions present themselves as spokespersons, and are involved in agenda-building for science. Individual scientists and groups develop new approaches (including holistic ones) to link up with SUSTAINABILITY. Being able to invoke SUSTAINABILITY mobilizes symbolic and financial resources. Even if it also involves one in the debates and controversies about the environment, global climate change, and issues of expertise and decision making generally.

Abstract sponsors create a space, and to some extent a protected space, for scientific research, and are thus part of the evolving social contract between science and society. They also play havoc with existing disciplinary distinctions. Just as academic disciplines could emerge and stabilize through the backgrounding of sponsors, the "return" of the sponsors (concrete and now also abstract) introduces dynamics leading to hybrid scientific communities and hybrid forums carrying new or at least modified ways of knowledge production. Patient associations in medical and health research would be one, and striking, example (Callon and Rabeharisoa 2003). Research supported by new sponsors like the Bill and Melissa

---

12 A similar phenomenon is the recent, and reluctant, acceptance by scientists of accountability, because "we're spending the taxpayers' money."

13 By now, USERS have become important as an ideographic category as well (Shove and Rip 2001).

Gates Foundation, or idiosyncratic "upstart philanthropists" like Fred Kavli,[14] is
not bound to existing categories.

The Existing Regime is Opening up

The whole constellation of spaces and sponsors and modes of knowledge production
which appeared to stabilize, and even lock-in, after the Second World War as the
regime of "Science, The Endless Frontier" now appears to open up. If Gibbons et al.
(1994) are right, a new regime – their Mode 2 – is upon us. Such a programmatic
claim is premature, but they (especially in Nowotny et al. 2001) do offer evidence that
the existing regime is evolving, and opening up to more interactions with society.[15]
So the first step is to map ongoing changes, and that is where the perspective I
outlined is useful.

Part of the dynamics derives from overall changes in our societies, which have
been diagnosed as 'reflexive modernization' (Beck et al. 1994). A key element of this
diagnosis is that institutions of modernity, including science, cannot continue as they
were used to. The heterogeneity that is encountered at the moment can be deplored
(by scientific establishments) as threatening science-as-we-know-it.[16] But it can also
be seen as an opportunity, because openness and variety allows renewal similar to
what happened in the melting pot of the European Renaissance. As Beck and Lau
phrased it: "what appears as 'decay' and de-structuration in the unquestioningly
accepted frame of reference of first modernity (and in this respect is bracketed
off and marginalized), is conceptualized and analysed as a moment of potential
re-structuration and re-conceptualization in the theoretical perspective of reflexive
modernisation" (Beck and Lau 2005, 552).

The dynamics are not just institutional. There is "new" natural history, i.e.
pattern recognition modes of knowledge production supported by ICT tools, GIS

---

14 Fred Kavli has started the Kavli Foundation which creates Kavli Institutes in basic
    areas of astrophysics, nanoscience and neuroscience, the fields that Kavli is interested
    in. The phrase "upstart philanthropist" to characterize Kavli was used in a news article
    by Michael D. Lemonick about Kavli, in *Time*, August 13, 2007, p. 44.

15 Others come up with similar diagnoses of opening up of what was closed/protected
    before, cf. "porous university" (De Boer et al. 2002).

16 Cf. Asher et al. 1995: In October 1994, "the world science leaders" met in Jerusalem. They
    were defensive, but prepared to defend the bastion of science. And: "if we do not measure
    ourselves, somebody else will - "upper management," the government, funding agencies,
    whoever - and they will probably do an even worse job of it."

(Geographical Information Systems) being one example (Rip 2002b), there is the advent of technosciences, and in general, the renewed importance of tinkering in the lab and in the field. This merges into making things – up to plants and animals – that (might) work, and having experimental infrastructures ('platforms') for research purposes, which can also be exploited for product development. This is very visible in the newly fashionable and "theory-poor" field of nanoscience and nanotechnology (Nordmann 2008).

There are meso-level developments as well, starting as responses to ongoing changes but then contributing to them in their own right. A key development is the increasing importance of the (new) category of 'strategic research', epistemically as well as institutionally. In the formulation of Irvine & Martin (1984) it clearly reflects a new division of labour between the quest for excellence and for relevance – and it may actually combine them.[17]

- Basic research carried out with the expectation
- that it will produce a broad base of knowledge
- likely to form the background
- to the solution of recognized current or future practical problems

This creates a new protected, but not necessarily closed, space.[18] Drawing on this, Centres of Excellence & Relevance are becoming a new and important institutional form, within universities and outside them. Their viability relates to the emergence of markets of strategic research (Rip 2002a, 2007).

Strategic research has now become pervasive, and science institutions adapt and evolve (Rip 2000b, Rip 2002a). A regime of Strategic Science might emerge, replacing – or grafted on – the regime of Science, The Endless Frontier, with master narratives of technoscientific promise (Felt et al. 2007, 21-29) and relevant expertise. It is carried by an alliance between politicians and science policy makers on the

---

17 Stokes (1997) showed the possibility of such a combination ("Pasteur's Quadrant"). His analysis, however, is a typology rather than diagnosis of dynamics. Cf. his use of historical figures (Pasteur, Bohr and Edison) to typify the three main quadrants, and his neglect of the fourth quadrant.

18 There is also the new category of 'translational research', now very visible in biomedical and pharmaceutical research, up to use in the Roadmap of the US National Institutes of Health (Atkinson-Grosjean 2006, 171). The category is more broadly applicable, e.g. to engineering sciences and to environmental sciences (where the 'translation' is towards decision making). The fact that there is such a category tells us something about changes and how these are captured with a label – under which a new protected space can function.

one hand, and a new elite of scientists promising to contribute to wealth creation and sustainability, on the other hand.

Part of the evolving regime, but more difficult to handle, is public scrutiny of science, ranging from accountability to involvement with publics. This is linked to increasing recognition of the value of experience-based knowledge, and further shifts in the notion of expertise (Callon et al. 2001 and 2009). At the same time, there are attempts at epistemic authority by those who used to be seen as outsiders, up to the US Congress pronouncing on what is 'sound science' (i.e. direct observation rather than models and 'theory' – which might be used to support environmental and other regulation). Following Brown (1996), one can call this conservative epistemic politics.

To turn this mapping of ongoing changes into a diagnosis, I return to my analysis of protected spaces as a functional requirement for science, at micro-, meso- and macro-levels. Protected spaces are essentially ambivalent: protection enables and nurtures, but also constrains, and might imprison. While the earlier regime is opening up, there is also closing down towards new institutionalizations, including new protected spaces.[19] Functional equivalents of the disciplines under Mode 1 might emerge. With this ambivalence in mind I will discuss, in the next two sections, changes (challenges) in knowledge production, and responses of scientific institutions to the opening up of the earlier regime.

Ambivalences of Opening up Institutionalized Knowledge Production

The opening-up of the earlier regime occurs in a variety of ways, but a key dynamic is the recognition of non-institutionalized knowledge production and use, which had been excluded, over the centuries, from the core business of science. This 'core' business is now getting further 'recontextualized', building on earlier layers of recontextualisation of science in society (e.g. strategic research programmes from the 1970s onward). This is visualized in the diagram below (Figure 2), together with examples of new boundary interactions.

---

19 The terminology of 'opening up' and 'closing down' is inspired by Stirling (2008).

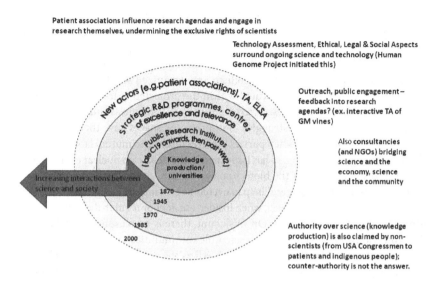

Patient associations influence research agendas and engage in
research themselves, undermining the exclusive rights of scientists

Technology Assessment, Ethical, Legal & Social Aspects
surround ongoing science and technology (Human
Genome Project initiated this)

Outreach, public engagement –
feedback into research
agendas? (ex. interactive TA of
GM vines)

Also consultancies
(and NGOs) bridging
science and the
economy, science
and the community

Authority over science (knowledge
production) is also claimed by non-
scientists (from USA Congressmen to
patients and indigenous people);
counter-authority is not the answer.

New actors (e.g. patient associations). TA, ELSA

Strategic R&D programmes, centres
of excellence and relevance

Public Research Institutes
(late C19 onwards, then post WW2)

Knowledge
production/
universities

Increasing interactions between
science and society

1870
1945
1970
1985
2000

Fig. 2   Opening-up and recontextualisation of science in society

For my overall argument, there is no need to discuss the details of the diagram (see Rip 2007). I note that all the new developments have a socio-epistemic character. In Nowotny et al. (2001), some of these are discussed as well, and linked to a notion of recontextualisation that is similar to the one I use here. Nowotny et al (2001) also introduce the notion of socially robust knowledge production, adding an additional societal layer to ongoing scientifically robust knowledge production. They see this as the way forward for science in society, but tend to argue the value of societal participation per se, i.e. a political consideration. There should be epistemic considerations as well. To introduce these, I will articulate the pursuit of robust knowledge as the central characteristic of 'scientific' knowledge production. Then, ongoing changes can be discussed as changes in the division of labour in the production of robust knowledge.

There are ambivalences involved in the production of robust knowledge, and not only because society enters the picture. Epistemically, robustness of knowledge has to do with how it will work again, at other times and in other places: it must be able to withstand variety and interference. To have robust outcomes, interactions and struggles are important, some of them (like peer review and struggle for visibility) focusing on traditional scientific robustness, while others will be linked to difference in values or interest strategies. The quality of the knowledge that is produced will improve through such agonistic (and sometimes antagonistic) interactions. These force actors to articulate the merits of their position, to search for arguments and counter-arguments. But they can also lead to impasses, or to repression of innovation.

This perspective on robust knowledge production is broader than the Popperian-Mertonian emphasis on fallibilism and organized scepticism. The latter now appears as a special case located within the protected space of an academic scientific community, and abstracted from many of the vicissitudes of the real world. Of course, agonistic struggles in unprotected spaces have their problems. They can lead to impasses, when parties limit themselves to mutual labeling of the other as contemptibly wrong, as has happened in the debate on nuclear energy (and happens to some extent in the biotechnology debate, although third parties e.g. supermarkets now intervene and help to overcome the impasse). But such processes occur in academic science as well, when insider-outsider or regular-deviant labeling hinders productive interaction. On this count, there is no reason to hark back to the protection afforded by academic scientific communities.

The production of robust knowledge as well as the assessment of its robustness need not be the exclusive domain of relevant scientific and technological communities. Epistemic quality should continue to be the goal, but it is not a matter of following methodological recipes. A key point, visible already in Kuhn (1977), is that knowledge production requires some closure of epistemic debate.[20] Such closure reduces complexity, but at a cost: alternatives will be backgrounded. Thus, there is a *prima facie* argument to entertain variety. But variety has costs as well: continued exploring need not lead to usable outcomes.

The ambivalence of entertaining variety is exacerbated when there are different cultural backgrounds, up to different "cosmovisions." An important domain where such struggles occur and have epistemic import is indigenous knowledge. By now, it is politically correct to accept claims from other cultural backgrounds to deserve a place under the epistemic sun,[21] but this creates tensions for Western science.[22]

---

20 The importance of provisional closure through reference to the status and productive use of expertise is also visible in regulatory science and now leads to attempts to create (even if precariously) new and authoritative forums.

21 Compare the advent of an "indigenous Renaissance" world-wide (Battiste 2000), and science funding agencies in countries like South-Africa and New Zealand supporting indigenous knowledge production.

22 Cf. how the International Council of Scientific Unions wrestled with the need to accommodate indigenous knowledge, also for political reasons, and wanted to continue to condemn pseudo-science. After some epistemological considerations, they offer an institutional-political answer: "Traditional knowledge is therefore neither intended to be in competition with science, nor is such a competition the necessary result of

It is now also practically correct to appreciate indigenous knowledge, as an as yet insufficiently tapped resource for development. Is it also epistemically correct? When cultural communities take over the quality control that used to be done by disciplinary communities there may be a problem of creating unproductive, because closed, protected spaces. The same exclusionary tactics will be involved ("we are the only ones who can judge") as Western scientists can apply, but these tactics will now foreground cultural heritage rather than new knowledge production.

A case in point is New Zealand's funding of Maori.Knowledge and Development research. It is but one example of the overall growing interest in indigenous knowledge, practically and politically, and the increasing voice and power of indigenous communities and non-Western approaches to knowledge (Battiste 2000, Smith 2000). In this case, the creation of a closed shop was encouraged by science funding actors emphasizing that Maori development research must be by Maori, for Maori, and follow a Maori world-view and approach to knowledge.[23] In other words, positive discrimination rather than exposing knowledge production to challenges so as to make it more robust. Of course, positive discrimination may be necessary for some

---

their interaction. On the contrary, as we have seen earlier, traditional knowledge has informed science from its very beginnings and it continues to do so until today. If a competition between science and traditional knowledge arises at all, then the initiative typically comes from people who want science to replace these other forms of knowledge. Pseudo-science, on the other hand, tries at least partly to delegitimize existing bodies of scientific knowledge by gaining equal epistemological status. The existence of pseudo-science as an enterprise fighting science is thus invariably bound to the existence of science whereas traditional knowledge stands on its own feet." (ICSU 2002, 11)

23 The phrase is from the (2001 and later) government science budgets, 'output class' Maori Knowledge and Development. The sentiment is carried broadly, as was clear in a June 2001 meeting of science officials and Maori representatives (http://www.morst.govt. nz/creating/maori/huiprogramme.html, accessed 10 October 2001). Pete Hodgson, Minister of Research, Science and Technology emphasized: "(..) in last year's Budget we created a new funding stream for research specifically by Maori for Maori. (..) The type of research supported by this stream embraces Maori customs and knowledge, using this base to research and develop tools and mechanisms to improve Maori health, social and economic well-being. In the same meeting, James Buwalda, CEO of the Ministry of Research, Science and Technology (MoRST), emphasized that indigenous knowledge systems have a valid role in economic, environmental and social development. And he adds: "Maori world-views have equal status alongside Western science." Similarly, Minister Hodgson was willing to say: "I think Maori think differently. (..) different ways to approach a problem, explore it and solve it. (..) good for us as a nation." Their embracing the Maori perspective marks a shift in policy at the top. In the meeting, called a *hui* to emphasize its link to Maori culture, Michael Walker (himself Maori) referred, somewhat cynically, to "the hymn sheet of the science/research agencies" about the importance of Maori knowledge. Hymn sheets may well have effects.

time, to nurture what still has to grow. The ambivalence returns with the question how long the nurturing should continue, and what its form should be.

The reference to "what still has to grow" may be found condescending, and the closed-circuit message of "by Maori, for Maori and with a Maori worldview" may be applicable to some peer review circuits in Western academic science as well. Still, it is important to keep the question of epistemic quality alive, also for indigenous knowledge, without it becoming an excuse to reject indigenous knowledge approaches out of hand. The key entrance point is the creation of spaces and how they function. The science policy initiative in New Zealand created a protected space at the meso-level, and could structure it only in terms of the way present funding agencies do their business. Thus, responsibility for the emergence of the closed shop will rest with them just as much as with emancipatory movements for indigenous knowledge.

Institutional Responses of Funding Agencies and Universities

The New Zealand science policy initiative is one example of science institutions being on the move, half-heartedly or actively engaging with the new challenges. In a sense, they are forced to move because their context is changing, and because they experience internal changes, e.g. new ways of knowledge production. I am talking here about the traditional institutions of science which are set in their ways, not about new types of institutions for whom opening up of the regime offers opportunities. However, for the question of overall change (and its diagnosis) the traditional institutions are a key entrance point because they cover a large part of the system of science. Thus, I will focus on traditional institutions, and offer an assessment of how funding agencies and universities are changing, and may change further.

Funding Agencies

From the 1950s onward, national-level science funding agencies have become a keystone institution for the modern social contract between science and society, as they were integrated in the inward-looking world of the Republic of Science (Polanyi 1962; Rip 1994). They have a strong institutional identity which continues to be reproduced even while circumstances are changing. As government agencies, they need to be accountable, so will not find it easy to respond flexibly to changing circumstances. In a sense, they are prisoners of their own achievement in doing a good job (i.e. funding research and assessing proposals).

Even so, the external pressures for relevance of science from the 1970s onwards had to be responded to, somehow. The funding agencies adapted, some more than others, [24] and shifted their practices by including relevance or merit criteria in addition to scientific quality of proposals. At first the additional criterion of relevance was not taken very seriously, and most often judged by scientists, and thus reabsorbed into the Republic of Science. Subsequently, the category of strategic research covered more and more of the research activities sponsored by funding agencies, and definitely featured in annual reports and strategic plans.

After the 1980s, links with market actors had become unavoidable, and were integrated (to variable extent) in their workings, e.g. in their consultations and in the composition of boards and panels. The need for broader consultation about strategies (and the need to articulate strategies at all) was visible already in the 1990s, and became generally accepted and widely practised in the 2000s. There is some opening up to plural stakeholders (not just market actors). By the late 2000s, overall changes in the science system were taking hold, including the more active role of patient associations and environmental groups, and the reference to "responsible development of science", in particular of newly emerging science and technology like nanotechnology (Kearnes and Rip 2009).

It is not clear whether these new developments will be temporary exercises, and be reabsorbed into the main thrust of the enlightened modernist response. [25] Given the strong mission of national-level funding agencies, and their need to be accountable, they cannot shift very much. If the ecology of the science system would change, however, for example because of the increasing importance of private funding bodies, especially charitable foundations, they could, and would have to move. [26] In the UK, where the Wellcome Trust funds more medical research than

---

24 For each national level funding agency (or agencies), the institutional path works out differently, with some resisting the pressure to include relevance (e.g. Germany), and others embracing it, at least in public declarations (e.g. UK).The German Deutsche Forschungsgemeinschaft justifies its reluctance by referring to the freedom of research, as laid down in the Constitution.

25 This terminology derives from an analysis of responses of science institutions to reflexive modernization (Delvenne and Rip 2009). One possibility then is that modernist approaches continue, but in an enlightened way.

26 I have drawn up scenarios of how funding agencies might develop, maintaining their core assets, but moving more freely in the ecology of national research systems (Rip 2000).

the government funding body (Medical Research Council), the move has started, there are joint programs and coordination.

Universities

The Elzinga & Wittrock (1984) volume on universities as a "home" (i.e. protected space) for the scientist, marks the beginning of an ever-expanding set of studies and comments on how universities are endangered by bureaucracy and "epistemic drift". Focal points are "new public management" as imposed on the universities from above (but often embraced by boards and administration as a way of strengthening their role as a "steering core" (cf. Clark 1998)), and the notion of a "third mission", towards society, which is felt as an imposition, even if some entrepreneurial universities take it as part of their profile (together with excellence).

This literature begs the question whether scientists should have a "home" (a protected space) "of their own", and whether that should be the university. Many universities, and most other institutions of higher education, are not dedicated to research. Also, there is proliferation of higher education institutions globally, up to claims to create research universities. Indian tycoon Anil Agarwal is building a university town, and was quoted in *Financial Times* (July 22, 2006) as saying: 'Vedanta University will be modelled on the likes of Harvard, Oxford and Stanford, catering for 100,000 students. 'What is money for if not to be made and given back to society?'

For my question about responses of universities, there are two interesting developments. First, attempts to create conglomerates. In the Netherlands, Wageningen University and Research Centre is a (precarious) combination of an agricultural university and dedicated agricultural research institutes. In France, there are collaborations between universities, Centre National des Recherches Scientifiques, and some of the big public research institutes. In South Africa, the alliance between the University of Pretoria and the Council for Scientific and Industrial Research has drawn attention. In Germany, Göttingen University has created an alliance with five Max-Planck-Institutes and other research institutions in the area. The establishment of Karlsruhe Institute of Technology, a merger between the university and the big public research centre in Karlsruhe, is a recent and very visible example. The message is that the traditional mission and boundary of the research university is not sacrosanct.

Second, the emergence of a new kind of entity, Centres of Excellence and Relevance embodying and pursuing strategic research. This started in the 1980s with the USA Engineering Research Centers, the UK Interdisciplinary Research Centres, and the Australian Collaborative Research Centres. By now, Centers of Excellence and Relevance emerge everywhere, and they are not limited to the context of research universities. In fact, they are a new species in the "ecology" of present research and innovation systems.

Such Centres can thrive because there is, by now, a 'market' for strategic research, as well as direct support of excellence by funding agencies and independent sponsors. When such Centres are part of a university, they are somewhat independent in terms of resource mobilization, and they can throw their weight around because they are important for the profile and competitive position of the university. In Rip (2002a), I have used my own university and its MESA+ Institute for Nanotechnology as a case study. Subsequent developments show the mutual dependency of the university and this Centre for Excellence and Relevance. To put it bluntly: the university is bursting at its seams because it houses such Centres. It has to re-invent itself – or give up being a research university.

The net effect is reinforcement of the pressure on research universities to transform themselves into the equivalent of a holding company, as is visualized in the diagram below (Figure 3). As soon as this happens, there will be openings for further developments, including the emergence (and design) of new kinds of protected spaces.

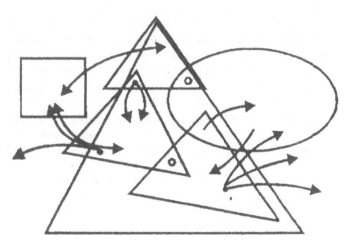

Fig. 3   The university 'complex' of the future

In Conclusion

The constellation of partially nested protected spaces of the regime of Science, The Endless Frontier, is opening up, and at all levels. Some such spaces, like funding agencies, modify themselves but essentially continue their path. Universities have more activities and concerns than protecting scientific research, but if they are research universities excellence and relevance of their research is an important part of their profile. Rather than continuing their traditional autonomy, they now enter into symbiotic arrangements. Centres of Excellence and Relevance are already somewhat independent of the university, and are becoming protected spaces in their own right. To coin a phrase, they could be the "home" of Mode 2 knowledge production.

The other main trend is the recognition of the value of experience-based knowledge. This has created openings, and there are experiments, but there are no institutionalized protected spaces yet. In the case of indigenous knowledge in countries like New Zealand where the issue is politically sensitive, incipient institutionalization showed the ambivalences of protected spaces (epistemic and institutional). There are other interesting developments as well, like the recognition of consultancies and environmental organizations as carriers of knowledge production.[27]

My socio-epistemic diagnosis (at the micro-level) of the need for protected spaces, even if their productivity is based on exclusion, is relevant at meso- and macro-levels as well. Again, there is the essential tension between entertaining variety (to ensure innovativeness) and maintaining some closure (to be productive). The risk is that these tensions will be short-circuited through institutionalization focusing on short-term productivity. Thus, in general, one should maintain (and even cherish) some heterogeneity, so as to avoid reducing complexity too much. To postpone a lock-in, one has to be prepared to live in (partially) unprotected spaces.

There is a governance aspect as well. This is brought out well in the MASIS Report (Markus et al. 2009), where US President Obama's phrase about "restoring science to its rightful place" is rephrased as a question about an *adequate* place of science in society, taking ongoing changes and contestations into account.

> [There is a] patchwork of transformations and tensions [which] does not result in a clear picture of an 'adequate' place of science in society. In fact, the open debate about the place of science in society should continue, and experiments to address tensions and other challenges should be welcomed.

---

27 Claudia Neubauer has called this the emergence of a "third sector" of knowledge production, in a paper for the preparation of the Expert Group Report *Taking the European Knowledge Society Seriously* (Felt et al. 2007).

The challenge is to support ongoing dynamics, rather than containing them, so dynamic governance is called for. (Markus et al. 2009, 4-5) [28]

Given the uncertain future of unprotected spaces and dynamic governance, the immediate challenge is to avoid reducing uncertainty by reifying an earlier epistemic pattern ("this is what science is, interference will not be tolerated), and/or an earlier institutional pattern ("let's continue with what we have been doing & organizing all along, perhaps modify it a little"). Of course, there are epistemic and institutional achievements that should be cherished, but even then, one has to understand how they came about, and whether they can continue to be productive under the new circumstances, or should be modified, even be replaced. This is how we can arrive at a dynamic diagnosis, profiting from a long-term perspective and considering the evolving ecology of science systems in context.

Bibliography

Asher, I., Keynan, A., & Zadok, M. (Eds.). (1995). *Strategies for the national support of basic research: An international comparison.* Jerusalem: The Israel Academy of Sciences and Humanities.

Atkinson-Grosjean, J. (2006). *Public science, private interests. Culture and commerce in Canada's networks of Centres of Excellence.* Toronto: University of Toronto Press.

Battiste, M. (Ed.). (2000). *Reclaiming indigenous voice and vision.* Vancouver and Toronto: UBC Press.

Beck, U., Giddens, A., & Lash, S. (1994). *Reflexive modernization.* Cambridge: Polity Press.

---

28 These quotes are part of a more specific diagnosis: "The revival of excellence of science as a goal, reinforced by the establishment of the European Research Council, provides an occasion for international competition, and for performance indicators based exclusively on publications in ISI-indexed journals. At the same, there are calls for increased democratization of science, concretely, the involvement of more stakeholders. More stakeholders, and existing stakeholders in new roles, are involved." "There are also developments in the governance of science in society. The governance of scientific institutions is under pressure, not least because of different contexts of governance, simultaneously pushing innovation, democratization and scientific integrity. New forms of governance are emerging: the discourse on responsible development, including attention to ethics and codes of conduct; interactive forms of technology assessment; and experiments with public engagement. Again, these are not without tensions, but they indicate that we do not have to fall back on traditional forms of governance." (Markus et al. 2009, 4-5)

Beck, U., & Lau, C. (2005). Second modernity as a research agenda: Theoretical and empirical explorations in the 'meta-change' of modern society. *British Journal of Sociology* *99*(4), 525-557.

Berwick, D. M. (2005). Broadening the view of evidence-based medicine. *Quality and Safety of Health Care, 14,* 315-1316.

Biagioli, M. (1993). *Galileo, Courtier. The practice of science in the culture of absolutism.* Chicago, IL: Chicago University Press.

Brown, G. E. (1996). *Environmental science under siege.* A Report by Repr. George E. Brown, Jr., US Congress, Oct. 23, 1996.

Brown, G. E. (1997). Environmental science under siege in the U.S. Congress. *Environment: Science and Policy for Sustainable Development, 39*(2), 12–31. https://doi. org/10.1080/00139159709604359

Bush, V. (1945). *Science -- The endless frontier. A report to the President on a program for postwar scientific research.* Washington, D.C.: July 1945. Reprinted, with appendices and a foreword by Daniel J. Kevles, by the National Science Foundation, Washington, D.C., 1990.

Bush, V., Kevles, D. J., & Bloch, E. (1990). *Science--the endless frontier: a report to the President on a program for postwar scientific research.* [Washington, D.C.: National Science Foundation. Retrieved from //catalog.hathitrust.org/Record/009239540

Callon, M., Lascoumes, P., & Barthe, Y. (2001). *Agir dans un monde incertain. Essai sur la démocratie technique.* Paris: Ed. du Seuil.

Callon, M., & Rabeharisoa, V. (2003). Research "in the wild" and the shaping of new social identities. *Technology in Society, 25,* 193-204.

Callon, M., Lascoumes, P., & Barthe, Y. (2009). *Acting in an uncertain world. An essay on technical democracy.* Cambridge, MA: MIT Press.

Campbell, D. T. (1979). A tribal model of the social system vehicle carrying scientific knowledge. *Knowledge, 1*(2), 181-201.

Burton R. C. (1998). *Creating entrepreneurial universities. Organizational pathways of transformation.* Oxford: Pergamon Press.

De Boer, H., Huisman, J., Klemperer, A., van der Meulen, B., Neave, G., Theisens, H., & van der Wende, M. (2002). *Academia in the 21st Century. An analysis of trends and perspectives in higher education and research.* The Hague: AWT (Adviesraad voor het Wetenschaps- en Technologiebeleid).

Delvenne, P., & Rip, A. (2009). Reflexive modernization in action: Pathways of science and technology institutions. Paper submitted to *Theory, Culture and Society.*

Eamon, W. (1985). From the secrets of nature to public knowledge: The origins of the concept of openness in science. *Minerva, 23*(3), 321-347.

Etzkowitz, H., & Leydesdorff, L. (2000). The dynamics of innovation: From national systems and "mode 2" to a triple helix of university-industry-government relations. *Research Policy, 29,* 109-123.

Felt, U., Europäische Kommission, & Europäische Kommission (Eds.). (2007). *Taking European knowledge society seriously: report of the Expert Group on Science and Governance to the Science, Economy and Society Directorate, Directorate-General for Research, European Commission.* Luxembourg: Off. for Official Publ. of the Europ. Communities.

Gibbons, M., Limoges, C., Nowotny, H., Schwartzman, S., Scott, P., & Trow, M. (1994). *The new production of knowledge. The dynamics of science and research in contemporary societies.* London: Sage.

Gilbert, G. N., & Mulkay, M. (1984). *Opening Pandora's Box. A sociological analysis of scientists' discourse*. Cambridge: Cambridge University Press.

Guston, D. H., & Kenniston, K. (1994). Introduction: The social contract for science. In D. H. Guston & K. Kenniston (Eds.), *The fragile contract. University science and the Federal Government* (pp. 1-41). Cambridge, MA: MIT Press.

Hacking, I. (1992). The self-vindication of the laboratory sciences. In A. Pickering (Ed.), *Science as practice and culture* (pp. 29-64). Chicago, IL: University of Chicago Press.

Hessels, L. K., & van Lente, H. (2008). Re-thinking new knowledge production: A literature review and a research agenda. *Research Policy,37*, 740-760.

Hufbauer, K. (1982). *The formation of the German chemical community (1720-1795)*. Berkeley, CA: University of California Press.

ICSU (2002). *Science and traditional knowledge. Report from the ICSU Study Group on science and traditional knowledge* (4th ed.). Paris: International Council of Scientific Unions.

Irvine, J., & Martin, B. R. (1984). *Foresight in science. Picking the winners*. London: Frances Pinter.

Kearnes, M., & Rip, A. (2009). The emerging governance landscape of nanotechnology. In S. Gammel, A. Losch, & A. Nordmann (Eds.), *Jenseits von Regulierung: Zum politischen Umgang mit Nanotechnologie*. Berlin: Akademische Verlagsanstalt, forthcoming.

Kogan, M. (2005). Modes of knowledge and patterns of power. *Higher Education, 49*, 9-30.

Kuhn, T. S. (1970). *The structure of scientific revolutions* (2nd ed.). Chicago, IL: University of Chicago Press.

Kuhn, T. S. (1977). *The essential tension. Selected studies in scientific tradition and change*. Chicago, IL: University of Chicago Press.

Latour, B., & Woolgar, S. (1979). *Laboratory life. The social construction of scientific fact*. Beverly Hills and London: Sage.

MacLeod, R. (1972). Resources of science in Victorian England: The endowment of science movement, 1868-1900. In P. Mathias (Ed.), *Science and Society, 1600-1900* (pp. 111-166). Cambridge: Cambridge University Press.

MASIS expert group. (2009). *Challenging futures of science in society: Emerging trends and cutting-edge issues*. Luxembourg: Publications Office of the European Union.

Mendelsohn, E. (1969). Three scientific revolutions. In P. J. Piccard (Ed.), *Science and policy issues. Lectures in government and science* (pp. 19-36). Ithasca, IL: F.E. Peacock Publishers.

Merton, R. K. (1974). *The sociology of science: theoretical and empirical investigations* (4th ed.). Chicago, IL: University of Chicago Press.

Morrell, J., & Thackray, A. (1981). *Gentlemen of science. Early years of the British Association for the Advancement of Science*. Oxford: Oxford University Press.

Nordmann, A. (2008). Philosophy of technoscience. In G. Schmid (Ed.), *Nanotechnology. Volume 1: Principles and fundamentals* (pp. 217-243). Weinheim: Wiley-VCH Verlag.

Nowotny, H., Scott, P., & Gibbons, M. (2001). *Re-thinking science. Knowledge and the public in an age of uncertainty*. Cambridge: Polity Press.

Polanyi, M. (1962). The republic of science. Its political and economic theory. *Minerva, 1*(1), 54-73.

Polanyi, M. (1963). The potential theory of adsorption. Authority in science has its uses and its dangers. *Science, 141*, 1010-1013.

Rip, A. (1982). The development of restrictedness in the sciences. In N. Elias, H. Martins & R. Whitley (Eds.), *Scientific establishments and hierarchies* (pp. 219-238). Dordrecht: Kluwer.

Rip, A. (1985). Commentary: Peer review is alive and well in the United States. *Science, Technology & Human Values, 10*(3), 82-86.

Rip, A. (1988). Contextual transformations in contemporary science. In A. Jamison (Ed.), *Keeping science straight. A critical look at the assessment of science and technology* (pp. 59-87). Gothenburg: Department of Theory of Science, University of Gothenburg.

Rip, A. (1994). The republic of science in the 1990s. *Higher Education, 28*, 3-23.

Rip, A. (1997). A cognitive approach to relevance of science. *Social Science Information, 36*(4), 615-640.

Rip, A. (2000a). Fashions, lock-Ins, and the heterogeneity of knowledge production. In M. Jacob & T. Hellstrom (Eds.), *The future of knowledge production in the academy* (pp. 28-39). Buckingham: Open University Press.

Rip, A. (2000b). Higher forms of nonsense. *European Review, 8*(4), 467-485.

Rip, A. (2002a). Regional innovation systems and the advent of strategic science. *Journal of Technology Transfer, 27*, 123-131.

Rip, A. (2002b). Science for the 21st Century. In P. Tindemans, A. Verrijn-Stuart & R. Visser (Eds.), *The future of the sciences and humanities. Four analytical essays and a critical debate on the future of scholastic endeavor* (pp. 99-148). Amsterdam: Amsterdam University Press.

Rip, A. (2007). Research choices and directions – in changing contexts. In M. Deblonde et al. (Eds.), *Nano researchers facing choices* (pp. 33-48). Antwerpen: Universitair Centrum Sint-Ignatius.

Rip, A., & van der Meulen, B. J. R. (1996). The post-modern research system. *Science & Public Policy, 23*(5), 343-352. Also published in R. Barre, M. Gibbons, Sir J. Maddox, B. Martin & P. Papon (Eds.), *Science in Tomorrow's Europe* (pp. 51-67). Paris: Economica International, 1997.

Rip, A., & van Steijn, F. A. J. (1985). *Effecten van de stimulering van de biotechnologie op de academische cultuur en de mogelijkheden tot kennisoverdracht.* Amsterdam: Dept. Science Dynamics. A Report to the Government Office of Science Policy. Also published in *Maatschappelijke aspecten van de biotechnologie* (pp. 169-190). Den Haag: Staatsuitgeverij.

Ross, S. (1962). Scientist: The story of a word. *Annals of Science, 18*, 65-85.

Schafer, W. (Ed.). (1983). *Finalization in science. The social orientation of scientific progress.* Dordrecht etc.: D. Reidel.

Shapin, S. (1994). *A social history of truth. Civility and science in seventeenth-century England.* Chicago and London: University of Chicago Press.

Shove, E., & Rip, A. (2000). Users and unicorns: a discussion of mythical beasts. *Science and Public Policy, 27*(3), 175-182.

Slack, J. (1972). Class struggle among the molecules. In T. Pateman (Ed.), *Countercourse* (pp. 202-217). Harmondsworth: Penguin.

Stirling, A. (2008). "Opening up" and "closing down". Power, participation and pluralism in the social appraisal of technology. *Science, Technology & Human Values, 33*(2), 262-294.

Stokes, D. (1997). *Pasteur's quadrant.* Washington, D.C.: Brookings Institution.

Toulmin, S. (1990). *Cosmopolis. The hidden agenda of modernity.* Chicago, IL: University of Chicago Press.

Van den Belt, H., & Rip, A. (1987). The Nelson-Winter/Dosi model and synthetic dyechemistry. In W. E. Bijker, T. P. Hughes & T. J. Pinch (Eds.), *The social construction of technological systems. New Directions in the sociology and history of technology* (pp. 135-158). Cambridge, MA: MIT Press.

Van den Daele, W. (1977a). Die soziale Konstruktion der Wissenschaft – Institutionalisierung und Definition der positiven Wissenschaft in der zweiten Hälfte des 17. Jahrhunderts. In G. Bohme, W. Van den Daele & W. Krohn (Eds.), *Experimentelle Philosophie. Ursprünge autonomer Wissenschaftsentwicklung* (pp. 129-182). Frankfurt a/Main: Suhrkamp.

Van den Daele, W. (1977b). The social construction of science: Institutionalization and definition of positive science in the latter half of the 17th Century. In E. Mendelsohn, P. Weingart & R. Whitley (Eds.), *The social production of scientific knowledge* (pp. 27-54). Dordrecht: Reidel.

Wittrock, B., & Elzinga, A. (Eds.). (1985). *The university research system. The public policies of the home of the scientists.* Stockholm: Almqvist & Wiksell International.

Chapter 2
Science Institutions and Grand Challenges of Society: A Scenario*

Another Grand Challenge

Science and technology can, and should, play a role in meeting "grand challenges" of society. But how, exactly?

One can dream, and reason back from the "dream" (the vision of a desired state of affairs) to stipulate ways to realize it. Alternatively, one can diagnose ongoing developments, project what might happen, and attempt to modulate the developments to do better, somehow. In practice, both happen all the time, in the small, when strategic decisions are made, and in the large, as when nation states or consortiums of nation states like the European Union, draw up visions of the future (for example, Korea's *Science and Technology Vision for the Future* (KISTEP etc 2010), the *Europe 2020* policy document (European Commission 2010)) and may refer to them when deciding on actions here and now.

How do science and technology come in? One sees reference to "grand challenges" of society in science policy documents and background studies. Foresight studies create a perspective, and this can mobilize people and institutions. There appear to be two main ways to link science and technology to grand challenges of society. For example, when Research Councils UK formulated main priorities, one type could be characterized as 'responsiveness to national needs', the other type as 'exploiting technoscientific opportunities'. The two types lead to different styles of implementation and ways of mobilizing institutions. This is clear already in the brief descriptions of the priorities (RCUK 2009).

*   Source: Arie Rip, Science Institutions and Grand Challenges of Society: A Scenario. *Asian Research Policy Journal*, 2(1) (2011) 1-9. This paper is the slightly revised version of my keynote address to the 3rd KISTEP Foresight Symposium, Challenges and Strategies of the Future, Seoul, 19 August 2010. I am grateful for the various comments I received, and for the help of Byoung Soo Kim (LISTEP) in preparing my keynote address. –

NanoScience through Engineering to Application	Ageing: life-long health and wellbeing
Nanotechnologies can revolutionise society. They offer the potential of disruptive step changes in electronic materials, optics, computing, and in the application of physical and chemical understanding (in combination with biology) to generate novel and innovative self-assembled systems. The field is maturing rapidly, with a trend towards ever more complex, integrated nanosystems and structures. It is estimated that by 2015 products incorporating nanotechnology will contribute US$1 trillion to the global economy, and that the UK has a 10 percent share of the current market. To focus the UK research effort we will work through a series of Grand Challenges. These will be developed in conjunction with researchers and users in areas of societal importance such as energy, environmental remediation, the digital economy, and healthcare. An interdisciplinary, stage-gate approach spanning basic research through to application will be used. This will include studies on risk governance, economics, and social implications	There is an unprecedented demographic change underway in the UK with the proportion of young people declining whilst that of older people is increasing. By 2051, 40 percent of the population will be over 50 and one in four over 65. There are considerable benefits to the UK of having an active and healthy older population with potential economic, social, and health gains associated with healthy ageing and reducing dependency in later life. Ageing research is a long standing priority area for the Research Councils. The Research Councils will develop a new interdisciplinary initiative (£486M, investment over the CSR period involving all seven Research Councils) which will provide substantial longer term funding for new interdisciplinary centres targeting themes of healthy ageing and factors over the whole life course that may be major determinants of health and well being in later life. Centres will be focused on specific research themes drawing on the interdisciplinary strengths of the Research Councils, such as Quality of Life, Physical Frailty and Ageing Brain.

One can stipulate what science institutions should do, and in this example from Research Councils UK, there is a quite direct link between priority setting and institutional arrangements. Even then, it is often not easy to realize the priorities because their implementation is refracted through the interests and own dynamics of the various research performing entities. When the priority is formulated as 'exploiting technoscientific opportunities', its implementation is easier (because it gives free rein to the promises that scientists come up with) than when 'responsiveness to national needs' is put upfront. Sometimes, the two seem to merge, as when investment in nanoscience & technology is presented as furthering global competitiveness of the nation. The underlying assumption of eventual economic performance of nanotechnology is still quite uncertain, however.

In a first step, I propose to shift the perspective from priorities (in relation to grand challenges of society) and their implementation, to the question what existing and future science and scientific institutions can <u>actually</u> achieve. They cannot just be "commanded" to achieve whatever goals are set higher-up (not even in Korea where there is this strong sense that everybody should work towards further development of the country). Apart from the fact that research results cannot be simply produced as desired (some things happen to be difficult or impossible), there are interests and own dynamics of these institutions, embedded in structures and processes in (national) systems of research and innovation. What can be achieved depends on these structures and processes, just as much as it depends on foresight and priority setting. (There is more to say, of course, about inertia of existing institutions, about industry structures and national specialization, and about political and institutional cultures in a country.)

Thus, there is another Grand Challenge: how to grow, maintain and keep productive a research and innovation system (national or otherwise). The Korean *Vision for the Future* (KISTEP etc 2010) discusses this in Part V, but does not offer much background analysis. This is a general phenomenon: concern about the research and innovation system and its productivity (along a number of dimensions) is quickly rephrased comparatively, and in terms of competition: are we doing as well as the Americans, or the Japanese? And the concern is further reduced to comparisons of indicators, rather than substantial diagnosis of structures and patterns and institutions in the system which shape what is no, produced, and may be produced in the future.

In general, while the challenge of institutional capacity building (including receptivity to grand challenges of society) is recognized, actual measures tend to be ad-hoc and superficial. Hobbyhorses of key people and dominant narratives (like the linear model of investing in science and technology to realize innovation to realize economic growth) continue to be more important in shaping what happens than in-depth analysis and diagnosis of what is happening and what future developments of the system could be. For the record, I add that existing scholarly literature and reports on national research and innovation systems often suffer from the same limitations (and tend to be very descriptive). More sophisticated approaches are being developed, however, particularly in Europe (see for example Smits and Kuhlmann 2004, and Schön et al. 2010).

In a second step, I will discuss the Grand Challenge of science institutions and national systems with the help of a foresight exercise. I will create a scenario about their further development which is informed by the more sophisticated approaches to science institutions (see Rip 2000, Rip 2011, and for the scenario approach in terms of 'endogenous futures' also Robinson 2009).

The scenario should be seen as a thought experiment, in which one can explore what might happen. In the thought experiment I present, part of my focus will be on trends that have perverse effects. Not because I want to paint a doom scenario, but to show how one can learn from a scenario exercise about ways to do better.[1]

A Scenario about Changes (up to Partial Collapse and Revival) in Science Institutions

I will present the scenario in three stages: the present, the near future, and the "cusp" of 2018 and its aftermath.

It is the year 2011, and the landscape of science institutions worldwide is under a variety of incentives and pressures.

Research universities write mission statements that emphasize excellence in research as well as their ability to serve society by contributing to innovation. They announce their intention to be in the top 100 of the Shanghai ranking, as the University of Göttingen in Germany did when submitting its bid to the German Federal Government's *Excellenz* Initiative, and the University of the Witwatersrand in South Africa did to show it wanted to recoup its status as an excellent research university. In Korea's *Vision for the Future* (p. 56), it is stated that there should be 10, rather than the present 2 universities among the top 100 universities. – The Shanghai list will become crowded …

This trend is pushed by governments and government agencies asking for indicators of performance, up to cultivating Nobel Prize winners (as the Korean *Vision for the Future* phrased it, p. 54). Scientists and science institutions respond by working to meet these indicators, somewhat independently of their actual, substantial performance.

At the same time, the promise of high technosciences like genomics, stemcells, nanotechnology, now also hydrogen economy, is pushed by (some) scientists and (most) government agencies alike. There are attempts to identify "key technolo-

---

1   In this respect, my scenario has the same function as the system-dynamics-based 1973 Report to the Club of Rome, which focused on limited global resources and environmental degradation, and created an early warning message. Korea's *Vision for the Future*, in its emphasis on green technology, takes up the message in a pro-active way. The focus on 'green technologies' is an attempt to mobilize science and technology to avoid the doom scenario – and create a globally competitive advantage for Korean industry at the same time. My scenario might help to identify further locally productive and globally competitive approaches.

gies" that a country (examples can be given of France and Germany) can invest in, in order to "pick winners". Somehow, every country seems to go for the same "winners", at least in high technosciences.

A key part of these dynamics are <u>funding races</u>, very visibly so for nanotechnology (see Figure 1). In spite of governments' interest in innovation, it is a funding race and not an innovation race. Firms tend to be reluctant to invest heavily in nanotechnology, because of uncertainties about eventual performance, as well as concerns about risks and about public acceptance. Consulting company Lux Research has now moved to focus on green nanotechnology as the way forward.

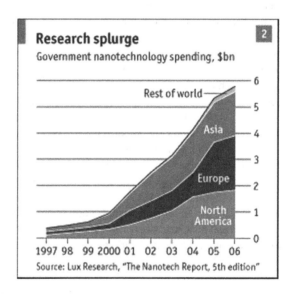

Fig. 1

This is not the whole story, however.

Research universities can include community orientation in their "third mission", and do so (and have to do so) in Latin America and Southern Africa. The lure of excellence remains strong, however.

At the side of innovation, there are pockets of user-driven and community-based innovation (compare Von Hippel 2004), the open-source movement, farmers' collectives etc where global competition is less important (see the analysis in Joly et al. 2010, who call this type of innovation 'collective experimentation').

There is a third development, where scientific knowledge production is opening up to accept input from other parties than professional scientists. Such inputs can come from stakeholders wanting to influence directions of research, or they can be invited by science funding agencies and other bodies to offer their views. The input can also be into knowledge production itself, as in health (contributions from patient associations), environment (consultancies and NGOs) and agriculture (farmers' collectives). Experience-based knowledge can then be integrated with professional-scientific knowledge production. As a trend, this overlaps with 'collective experimentation' as a mode of innovation.

These recent developments are still on the margins of the dominant narrative about excellence and the promises of high technosciences, however. Another opening up that is not at the margins anymore, is the question of scientific (or research) integrity. Even if fraud and plagiarism etc are often treated as a matter of individual failure rather than a problem of the system, there are system-level initiatives. The USA continues to have an Office of Research Integrity, the European Science Foundation has a working party on scientific integrity, and there are proposals for Codes of Conduct by established science institutions like Academies of Science.

A further step in this direction is that epistemological debate about research findings, methodologies and background perspectives is not limited to professional scientists anymore. Committees of the USA Congress are willing to expound on what is "sound science" – which for them excludes theoretical speculation and modeling (as in climate change research) in favour of empirical research, i.e. collecting "hard" data. From a different starting point, but with the same thrust (scientists need not have a monopoly on what is good science), indigenous and community knowledge and perspectives are pushed as real alternatives – and sometimes taken up as such, as in a New Zealand government Maori-oriented research funding program (see Rip 2002).

The variegated developments I sketched here all attempt to open up scientific institutions to broader considerations and broader inputs. In Figure 2 below, I position them as a next step in recontextualizing modern science (since the late 19th century – since 1870, to be precise). Earlier recontextualizations led to new public research institutes and subsequently, strategic research (funding) programs, and Centres for Excellence and Relevance.

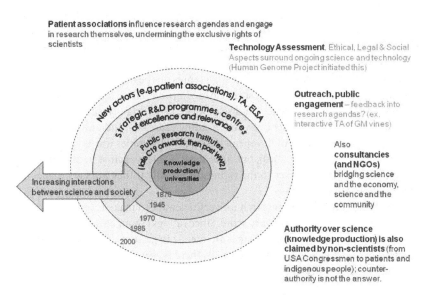

Fig. 2   Increasing re-contextualization of science in society

Given this evolving situation, with scientific institutions pursuing their own and often short-term interests, and responses from various societal actors and audiences, what might happen in the next 3 to 6 years?

Three main dynamics were outlined for the present: excellent research, science for innovation (mainly high technoscience), and opening up to new stakeholders. The further dynamics will be influenced by contexts and circumstances, of course, but I will abstract from such contingencies for this scenario exercise. And contingencies, even if major in their own right, may not have a large impact on the evolution of science institutions. For example, the present financial and economic crisis will have effects on spending on science and technology, but there is also a strong feeling that investments in science and technology will help overcome the crisis, at least in a long term perspective. More important are circumstances that are incited by the present dynamics, for example disappointments about what science and technology actually deliver, and thus a tendency to be more realistic about the promises of new science and technology.

The present patchwork has three strands that are important for the scenario: (1) the focus on indicators, which gives rise to a "derivatives" industry; (2) the attempts to link high technoscience with actual innovation and uptake; (3) some de-professionalization of science.

The first strand is dramatically visible in how the Shanghai ratings for universities have become a "derivative", foregrounding indicators rather than substantial excellence. A trade in means to achieve high scores on the indicators has emerged. As on the stock exchange, there is strategic buying and selling (of top researchers, of creating centers of excellence (and relevance), of setting priorities that follow the fashion). All this while basic research funding decreases, and universities have fewer margins. The League of European Research Universities warns against such a <u>reputation race</u> (Boulton 2010), but finds itself helpless against it, also because its members continue to refer to the ratings (especially if they are favourable) in public statements.

More and more aspiring research universities want to make it to the ranking. Thus, a demand emerges for ways to score. Since ratings (Shanghai and others) take awards and prizes into account, there is an interest in getting such prizes, but also to have more such prizes awarded. Actually, there is a supply: various bodies and rich individuals (the Kavli Prize is an example) start giving awards for excellence. In such a situation the tongue-in-cheek proposal on a blog of a scientist to have the Nobel Prize Committee award more prizes (there's much more research done now compared with fifty or one hundred years ago, so also more occasions to recognize and reward excellence) becomes a topic of real debate, up to pressures exerted on the Nobel Prize Committee to do something about the proposal.

The other race continues as well. Even if the focus of the promises might change, e.g. from nanotechnology to synthetic biology (as happens in the US), the pattern remains the same. Investments are argued in terms of global competition, and phrased as a race: we have to be first! Scientists go along: "high tech means high funding", whether it is actually useful or not. While there is increasing recognition that the promises are not easy to realize, the <u>promise race</u> continues. Like the arms race, it has a dynamic of its own. Even with the financial crisis, high technosciences continue to get funding, because it holds out hope of recovery.[2]

At the same time, there is increasing reluctance to buy into the promises as such. Business firms which have to deliver a product in the end had been reluctant all along (up to the point of being myopic, at least prone to waiting games, cf. Parandian et al. 2010). A pro-active approach emerges, somewhat independent of the overall promise

---

2   There is some sense in that: counter-cyclical investing. But this requires courage, and a
    feeling for what is, or may be, important. When decisions are predicated on a competition
    logic, however: doing better than others, and thus create wealth, somehow, they may be
    self-defeating, because everybody converges on the same priorities.

race. The gap between techno-scientific promises and actual products/services can be bridged by 'translational research'. The notion was used in drug development to bridge gap between proof of principle (of the working of a drug) and the actual formulation and administering of a drug in practice, and is now generalized to be a necessary complement to all promising techno-scientific options. New sponsors like the Bill & Melissa Gates Foundation start to emphasize translational research (without necessarily using the term) in addition to pursuing promises.

Thus, the 'market' for strategic research (Rip 2004) evolves further, now including translational research. Public research institutes, company R&D units, and some university departments become regular suppliers. It is not a classical market with independent suppliers and customers. The European Technology Platforms, pursuing anticipatory strategic coordination between research, uptake and projected use, become increasingly important as intermediaries, and in other regions of the world there are attempts to follow their example, or graft it on own ways of anticipatory coordination (as in Japan and South-Korea).

Two further, and in a sense complementary, movements occured with respect to the promise dynamics. Funding agencies in Europe started to require extended impact assessment (this is European Union terminology; the US National Science Foundation speaks of 'broader impacts criterion') as part of research proposals. At first, it remained limited to a few additional paragraphs in research proposals. But the logic of competition (for funding) kicked in, and proposal writers knowing their score could be increased by having a good impact statement started to hire consultants to write up such statements. The work of the experts assessing the proposals became more difficult. After a few years of learning by trial and error, the situation stabilized, and included links to translational research.

The fact that promising new science and technology also led to concerns, up to resistance (as with green biotechnology) had been a reason to include ELSA (research on Ethical, Social and Legal Aspects of the new science/technology) in funding programs, already in the 1990s for genomics (Rip 2009). This also happened with nanotechnology, but was now placed in the broader framework of 'responsible innovation' (or 'responsible development'). The reference to 'responsible' creates openings for more stakeholders, and for civil society generally, to be involved (Rip 2010). This remained tentative, however, even while there were interesting examples in some countries that might be taken up elsewhere.

The opening up that is occurring is still predicated on high technosciences, which had been a world onto itself and is now being contextualized. If one does not start from this world, a different picture emerges. There is collective experimentation (Joly et al. 2010) going on, which can take up high technoscience, but on terms set by the collective experimenters. High tech innovation in professional and amateur sports is a good example, but also "technoblending" in developing countries, when local technologies and experiences are combined with imported technology in a productive way.

A different picture also emerges in the health sector. The constructive role of patient associations continued. This was reinforced by a reconsideration of the high-science approach. As Berwick (2005) phrased it: "The movement for evidence-based medicine (..) transcended and improved upon local, experience-based knowledge, but has now "overshot the mark" and "excludes too much of the knowledge and practice that can be harvested from experience". While the opening-up that is occurring in a variety of ways may signal de-professionalization, the question of quality control (and quality assurance) has to be addressed even if traditional professionalized science is not the only arbiter anymore. Actually, re-professionalization occurs, as in the integration of Chinese traditional medicine and so-called Western medicine, in China, and also in Australia.

Thus, the evolving patchwork has recognizable strands, but there is no overall integration. Maybe one should not ask for overall integration, just go for "muddling through" (as Charles Lindblom phrased it, cf. Lindblom and Woodhouse 1993). But the "muddling through" might be hampered by the distance between excellence and high technoscience promises on the one hand, and the ongoing activities in research, product development, and collective experimentation on the other hand.

The "cusp" of 2018 and its aftermath.

The patchwork of developments might have continued in its amorphous and stumbling ways, were it not for a triggering event and its repercussions.

Remember the pressures on the Nobel Prize Committee to do more? When the 2016 Nobel Prizes were awarded in November of that year, the Committee also announced it would start awarding Nobel Prizes every half year, rather than only once every year. Subsequently, others felt free to start awarding prizes, or increase the frequency of prizes they were awarding already. After the first enthusiasm from the side of researchers and science institutions, there was a realization that the proliferation of prizes reduced their exclusive value. The various rating lists of universities which, because of the dynamics of the reputation race, had continued despite the criticisms of their methodologies, now lost their legitimacy: researchers were not interested anymore, and government bodies and funding agencies stopped taking such indicators of excellence into account. By 2018, the reputation race lost its thrust – this bubble of "derivatives" had burst.

Research universities were at a loss what to do now. Also, the groundswell of other kinds of universities (higher education institutions) increasing in numbers and variety could now be recognized for what was happening and what it might imply. Dedicated niche universities and colleges (up to indigenous universities in Latin America, and universities set up by millionaires in India) suddenly became visible as legitimate. Some of them followed the pattern of US community colleges, with no research, others were doing research, but often outside established research categories. Diversity was recognized as important.

One "bursted bubble" led to another. The dynamics of high technoscience promises, somewhat eroded already because of the interest in translational research, were now turned around: promises had to be accompanied by specifications how to achieve them. The license for doing excellent research now became: if no (attempt at) translational research, no support/funding for basic research.

Some institutions (new universities, public labs, strategic research programmes and centres of excellence and relevance) embraced this, because it justified their existence and thrust. Other institutions, in particular traditional research universities, accommodated so as to survive.

The shift towards realism when promising high technoscience made the original narrative less dominant, and thus allowed recognition of the value of other approaches (versions of collective experimentation). These had been important in practice already, but now became part of the official agenda.

All this worked out differently in different countries. Techno-scientific promises remained important in China and other Asian countries, although some, like South Korea, considered embedding in society as well. Globally, the net effect was that the importance of technoscientific promises in decision making about funding and about the role of research in meeting grand challenges became less, and there was more attention to what could actually be done in the short term.

There were protests about such short-termism. While many of these protests were just attempts to recover the freedom of scientists to go for excellence and for high technoscientific promises, the issue of short-termism remained important. Over time, the pendulum might swing back.

In Conclusion

Different types of conclusions are in order.

One is about the starting point, how to meet grand challenges. The basic message is: they were not met at first, because excellence and promising games were dominant. When the bubbles had burst, more concrete linkages were up front, but still as a patchwork, with no assurance of results. Maybe the whole idea of grand challenges should be seen as a mobiliser, enabling further work, rather than something that can and should be achieved.

Another conclusion is about science institutions. There is inertia of institutions (universities, funding agencies, government ministries). And there are the self-inflicted constraints of participating in reputation and promise races. Even when such patterns are recognized for what they are, institutions cannot easily step out of them. Only when the perverse effects of such patterns come home, they are able to take action.

At the system level, the important point is about diversity. Many well-intended policies may reduce diversity, and thus incentives for creativity and productivity. Reputation and promise races also push convergence rather than diversity. There are counter-currents (collective experimentation, non-professional knowledge production), but they will not come into their own unless the bubbles burst. Perhaps (and ironically) one should push the bubble so as to make it burst earlier, before it creates too much damage.

This then has implications for foresight, and in two ways: the methodology of foresight, and the role of foresighters.

Foresight has to take non-linearity of developments into account, where actions and reactions determine what happens, rather than some overall trend (cf. Robinson 2009). The collapse of the dynamics of ranking (as a derivative) is an interesting possible future. It is a thought experiment and need not happen that way. But it has a certain plausibility to it: things might develop this way in our kind of world.

One need not agree to the specifics of the scenario I presented here, but the methodological message is still important: there can be, and will be, non-linear developments. Which require scenario exercises rather than trends and roadmapping.

Foresighters are not just desk researchers who write a report. They are part (even if only a small part) of the dynamics they attempt to describe and diagnose. My own work on the future of research funding agencies is a case in point (Rip 2000), because it developed from interactions with funding agencies, and funding agency staff (in various countries) continue to invite me as a commentator. Foresight exercises like the Korean Vision for 2040 are more distant from present practices. But also there, one can envisage interaction with various actors who are making choices and decisions.

Bibliography

Berwick, D. M. (2005). Broadening the view of evidence-based medicine. *Quality and Safety of Health Care, 14*(5), 315-316.

LERU (2010). *University rankings: Diversity, excellence and the European initiative* (LERU Advice Paper #3, June 2010). Leuven: G. Boulton.

European Commission (2010). *Europe 2020. A European strategy for smart, sustainable and inclusive growth* (March 2010). Retrieved from http://europa.eu/press_room/pdf/complet_en_barroso___007_-_europe_2020_-_en_version.pdf

Joly, P.-B., Rip, A., & Callon, M. (2010). Reinventing innovation. In M. Arentsen, W. van Rossum & B. Steenge (Eds.), *Governance of innovation* (pp. 19-32). Cheltenham: Edward Elgar.

KISTEP, National Science & Technology Council, Ministry of Education, Science and Technology (2010). *Korea's dream and challenge. Science and technology vision for the future. Toward the year of 2040* (Draft, May 2010).

Lindblom, C. E., & Woodhouse, E. J. (1993). *The policy making process* (3rd ed.). Englewood Cliffs, NJ: Prentice Hall.

Parandian, A., Rip, A., & te Kulve, H. (2010). *Dual dynamics of technological promises and waiting games around nanotechnology.* Paper presented to the IGS & EU-SPRI Conference on Tentative Governance in Emerging Science and Technology. Enschede: University of Twente.

RCUK (2009). *RCUK Delivery Plan 2008/09 to 2010/11.* Retrieved from http://www.rcuk.ac.uk/documents/publications/anndeliveryplanrep2008-09.pdf

Rip, A. (2000). Higher forms of nonsense. *European Review, 8*(4), 467-485.

Rip, A. (2002). Science for the 21st Century. In P. Tindemans, A. Verrijn-Stuart & R. Visser (Eds.), *The future of the sciences and humanities. Four analytical essays and a critical debate on the future of scholastic endeavour* (pp. 99-148). Amsterdam: Amsterdam University Press.

Rip, A. (2004). Strategic research, post-modern universities and research training. *Higher Education Policy, 17,* 153-166.

Rip, A. (2009). Futures of ELSA. *EMBO Reports, 10*(7), 666-670.

Rip, A. (2010). De facto governance of nanotechnologies. In M. Goodwin, B.-J. Koops & R. Leenes (Eds.), *Dimensions of technology regulation* (pp. 285-308). Nijmegen: Wolf Legal Publishers.

Rip, A. (2011). Protected spaces of science: their emergence and further evolution in a changing world. In M. Carrier & A. Nordmann (Eds.), *Science in the context of application: Methodological change, conceptual transformation, cultural reorientation* (pp. 197-220). Dordrecht: Springer.

Robinson, D. K. R. (2009). Co-evolutionary scenarios: An application to prospecting futures of the responsible development of nanotechnology. *Technological Forecasting & Social Change, 76,* 1222-1239.

Schoen, A., Könnölä, T., Warnke, P., Barré, R., & Kuhlmann, S. (2011). Tailoring Foresight to field specificities. *Futures, 43*(3), 232–242. https://doi.org/10.1016/j.futures.2010.11.002

Smits, R., & Kuhlmann, S. (2004). The rise of systemic instruments in innovation policy. *International Journal of Foresight and Innovation Policy, 1*(1/2), 4-32.

Von Hippel, E. (2004). *Democratizing Innovation.* Cambridge, MA: MIT Press.

Chapter 3
Processes of Technological Innovation in Context – and Their Modulation*

Technological innovation in context has been studied by economists and sociologists of technical change and innovation. I shall present the insights and perspectives from this body of literature (including some of my own work), in order to highlight the dynamics of technological innovation processes and the possibilities to influence them – by managers, as well as governmental and societal actors. These actors often work with a limited view of the complexities of technological change and innovation, and they might do better if they were to use recent insights, as I have argued previously (Rip, 1995). Thus, a further topic, visible between the lines of my main exposé, is the relation between the 'theory' – i.e. insights from social scientific studies – and the 'practice' of policy or action.

More than thirty years ago, the American economists Nelson and Winter published an article about what a 'useful theory of innovation' might be, and they sketched an evolutionary approach: new technological options are introduced all the time but only some of the firms doing so would grow, depending on selection environments (Nelson and Winter, 1977). The late 1970s and early 1980s was also the time when national governments started to articulate technology and innovation policies. The Netherlands White Paper on Innovation (1979) actually used the Nelson and Winter distinction between technological trajectories and selection environments to categorize and, to some extent, derive policy measures and policy tools: stimulating variation, supporting selection, and improving interaction between variation and selection. Thus, theory was applied by policy makers, albeit in simplified form.

Why go back to the late 1970s? One reason is that the Nelson and Winter article was a seminal publication, and later work continues to refer to it. The other reason is that the example of the Netherlands White Paper shows interesting links between

---

*   Source: Chris Steyart and Bart van Looy (eds), Relational Practices, Participative Organizing. Bingley, UK: Emerald, 2010. Advanced Series in Management.

an understanding of the dynamics of technological innovation and the identification of opportunities for intervening; in this example, through policy measures.
Intervention requires simplification, not just the (hopefully productive) reduction
of complexity that any theory entails but also translation of projected goals into
concrete attempts to make a difference, for which theory can be mobilized.

In doing such translations, policy making draws on a modernist philosophy,
almost by definition. One indicator is the strong 'instrumentalist' push for robust
methods that allow the policy maker to make a difference, to exert influence – in
other words, act at a distance. The subtitle of Pressman and Wildavsky's (1984)
book, 'How Great Expectations in Washington Are Dashed in Oakland', summarizes both the modernist thrust and its limitations. The instrumentalist thrust may
already be counterproductive on its own terms, when its neglect of the complexities
and the own dynamics of the world out there in Oakland reduces the chance of
achieving its objectives.

Still, modernist approaches continue. This is linked to strong presumptions about
a envisaged better world, where the world-as-it-is appears to be an obstacle to be
overcome by dedicated action. But, if we live in a non-linear world where multi-actor
interactions and their sometimes unpredictable outcomes determine what happens,
such an approach, predicated on the idea of an actor working to realize a goal and
achieving it because of his/her efforts, will by definition be unproductive. I am not
implying that individual policy makers are never sensitive to the limited scope of
their action among a multitude of actions and interactions. The structure and culture
of the policy environment forces a modernist approach upon them, whether they
identify with it or not. (Rip, 1998)

The modernist philosophy highlights the heroism of the policy actor vis-à-vis
the system. But, there is a variety of actors and roles, and eventually a 'distributed
coherence' for which no single actor is responsible. Some actors may contribute
more to such self-organization than others but there is no general rule. If dominant
positions occur, these cannot be taken for granted and are, in that sense, contingent.
Instead of steering from a position of strength (which includes authority), there
are mutual translations. The policy actor is one of the actors – no more, no less.
But the policy actor's link with the *res publica* does create a special responsibility.
Even if a policy actor cannot do much more than induce or modulate ongoing
self-organization, s/he can be held accountable. And the 'shadow of authority' that
goes with the position can have effects, even if these are not linearly derived from
the authoritative role.

While I focussed on policy making and implementing, my point about modernist
approaches is general. Managers of different ilk are also modernists – although it is
easier for them to become reflexive and even turn to post-modern approaches because

they are much closer to ongoing processes (Visscher & Rip, 2003). 'Post-modern' is probably too strong a term, though.

In general, for policy actors and for managers, productive intervention, in the large and in the small, remains a challenge but not in a simple modernist fashion. A few authors have offered relevant perspectives and conceptualisations, such as analysis of transitions from 'fluid' to 'specific' situations (Garud, 1994) or from 'hot' to 'cold' situations (Callon, 2002). Evolutionary approaches are important because these emphasize the contingent character of eventual stabilisations. Relational approaches are another way to thematize non-modern management approaches. To intervene in fluid and distributed processes should be seen as a matter of modulation (Rip, 2006a). Brute change efforts are an extreme version of modulation and one that can be productive only in extreme circumstances such as a war economy.

In this chapter, I shall sketch the general perspective, based on an evolutionary approach to technological change, then introduce the notion of an innovation journey (after Van de Ven et al., 1999), identify specific opportunities for modulating ongoing processes, and in the conclusion come back to the role of government.

The Quest for Understanding: An Evolutionary Approach

Why go for evolutionary theory? Evolutionary economics and the related sociology of technological change continues to be of great interest to policy makers but is not really accepted by mainstream economics. In my view, this reflects less on the quality of the evolutionary approaches than on the limitations of mainstream economics itself, which is too far removed from what happens in the real world (at least, with respect to technological innovation). Economics has always had problems treating dynamic situations, and technological innovation is dynamic almost by definition.

The need for evolutionary theory derives from the recognition that technological innovation always implies a change with respect to what existed and, to some extent, a break with what was usual. In other words, a new technological option is a 'novelty' just like a mutation in Darwinian or other biological evolutionary theory. And like a mutation, the new option might well start out as a 'hopeful monstrosity' (Mokyr, 1990; Stoelhorst, 1997) that must be nurtured to improve, grow and survive the harsh selection environments – sometimes, the organization in which the novelty arises, is already a harsh selection environment.

Such an evolutionary theory is directly relevant to the issues of policy making and management. Its relevance goes deeper than policy makers using such insights to articulate policies they had defined for their own reasons (as happened in the

1979 Dutch White Paper on Innovation, which I discussed in the introduction). Policy makers and managers are themselves part of the evolutionary process. Thus, versions of strong planning and command-and-control approaches, or the equally ideological neo-classical economists' approach of leaving everything to the market except for a few generic fiscal measures to overcome collective-good problems, must be evaluated in terms of their function in specific evolutionary processes. Already, with Nelson and Winter (1977), the 'selection environment' of their theory is much broader than the market and includes the patent system and other institutional configurations. Intervention in such broader configurations and their evolution is not a simple matter. As I intimated already, the general approach is to understand ongoing dynamics and try to modulate them – this includes joint-learning approaches.

A further reason to prefer an evolutionary theory of technological innovation is the fact that the success of an innovation does not just derive from the power of the technological option on which it is based. I will develop this point, drawing on the insights from studies of processes of technological innovation. This will enable me to present an evolutionary theory of technological change and innovation in three steps.

The first step is to consider the introduction of novelty into an existing order. The analysis provided by Abernathy & Clark (1985) is a useful entrance point but must be extended. They emphasize how existing linkages – with customers and in markets – are broken, and how existing competencies (technological and other relevant capabilities within the firm) become irrelevant ('obsolete' is their term). Such 'de-alignment' is accompanied by 're-alignment': the building up of new linkages and competencies. One can think of firms having to work with new suppliers and customers, and having to retrain personnel. 'Re-alignment' creates links with the selection environment and thus some fit but this remains precarious. There is no guarantee of success and, in that sense, the evolutionary notion of a partially independent selection environment continues to apply.

Abernathy & Clark focus on the firm and its immediate business environment. But in fact, and especially if one considers longer-term developments and broader notions of success than short-term profits of a firm, societal linkages and competencies are involved as well (Rip, 1995). New products can be taken up in other sectors; think of information and communication technologies as an important example. Analysts like Freeman and Soete (1997) have pointed out that the necessary 're-alignments' within and across sectors take a long time. For them, this explains the so-called productivity paradox (investments in the use of ICT are not yet reflected in increased productivity). Longer-term changes can add up to transitions, and the patterns in such transitions can be understood in terms of re-alignments and their partial stabilization (Geels, 2005).

Another element in societal linkages and competencies is domestication of new technologies, including issues of public acceptance. For biotechnology and genomics, these issues are important in many countries, and managers may consider adding 'societal embedment' to their product development strategies (Deuten et al., 1997). With the emergence of nanotechnology, scientists, industrialists and policy makers now anticipate such issues and develop activities to avoid or mitigate them (Rip, 2006b). My theoretical point is that the de-alignment that occurs with respect to existing sectors and their cultures, and is expected and now discussed at the level of society as well, is not a matter of undue reluctance, or even resistance, by publics but a general feature of technological change. If technology promoters seek to impose re-alignment without recognizing the actual dynamics for what they are, their actions may well be counterproductive, as the impasse over agricultural ('green') biotechnology has shown.

In this first step, I have already gone further than the simple evolutionary approach of considering the introduction of novelty as a mutation or variation, and see selection mechanisms of various kinds as determining whether it survives and grows. In Nelson & Winter's theory and their subsequent model (Nelson and Winter, 1982), routines within firms to produce innovations are the retention mechanism (the third key component in any evolutionary explanation), and selection works on the firms in markets and other selection environments. Important also for my analysis is how Rip and Kemp (1998), and in a slightly different way Basalla (1988) and Levinthal (1998), have focused on new technological options rather than actual products, so that one can include the quasi-markets that have emerged for them (cf. Schaeffer, 1998) in the theory. This links up with earlier analyses of expectations and promises about new technological options (Van Lente, 1993; Van Lente & Rip, 1998), and recent studies as in Brown et al. (2000) and Borup et al. (2006).

This enables the next step in the theory, the recognition that variation is not random but shaped by anticipations, which can become stabilized as regimes or paradigms (Dosi, 1982). This links up with the phenomenon of dominant designs as, for example, the VHS video cassette recorder, which became dominant after a struggle with alternative, competing designs (Cusumano et al., 1997). Further innovation then occurs within the framework of the dominant design. Over time, the dominant design may shift or even be undermined and disappear. In the short run, innovations running counter to the dominant design and, more generally, to the rules of the dominant regime will be more difficult to realize, and they may well face barriers that are not easy to overcome.

These last observations indicate that the evolutionary perspective overlaps with what in general sociology is called institutionalization. Some novelties, at first precarious and surviving because of the promises claimed for them (up to the hype surrounding e-commerce, and now genomics and nanotechnology), will grow and become established – which is to say, they will become irreversible. Like institutions more generally, they will then resist further change. This occurs in the small, on the micro-level, as when organisations resist renewal from within or from without. But it also takes place in the large, on the meso-level of dominant designs, industry structures and standards, and on the macro-level of techno-economic paradigms. See, for instance, how Freeman and Louçã (2001), in their analysis of long waves in the economy of modern societies, highlight the role and dynamics of structural readjustment.

The third step is to open up this picture of a tendency to stabilization and insti-tutionalization, and to do so in two ways. There are broader developments that may unsettle stabilization. And there is the challenge of anticipation in an uncertain and largely unpredictable world. Actors develop ways to handle uncertainty – in evolutionary terms, by maintaining some variety and/or attempting to protect novelties against harsh selection. Other (often *de facto* or pattern) strategies are to reduce complexity by relying on shared expectations about future developments. Such shared expectations offer a sense of direction but that may be deceptive. A more sophisticated understanding of dynamics is important, starting with the recognition of non-linearity of developments.

Over time, there are shifts in functionalities and uses of an innovation, and there is uptake in other sectors, which induces further developments. The functions the telephone now fulfils are very different from the ones envisaged originally: business communication between the centre and the periphery of the town, and piping concerts from the concert hall in the city centre to the suburbs. Similarly, the idea of mobile telephony was raised as far back as the 1920s and led to prototypes, e.g. a motorcar with a radio, which allowed a link-up with receivers of the signal elsewhere – a far cry from present-day mobile telephony.

In other words, the strategies of managers – and, at one remove, policy makers – should not just be based on performance in present contexts, and actors' intentions and their predictions of eventual achievements. One way to do better is to develop scenarios that encompass non-linearity, for example, by drawing on a 'branching' model of technological development. Another way is to consider evolving contexts, i.e. shifts in the selection environments (markets and institutions), as part of the dynamics. This implies that one should speak of co-evolution rather than evolution. Nelson (1994) has already emphasized co-evolution of technology, industry structure and supporting institutions. We have used a broadly co-evolutionary approach to analyze and characterize processes where whole regimes were transformed, as from sailing ships to steamships (Geels, 2005). An important feature of co-evolution is its multi-level characteristic, where sequences of events at the micro-level, institutions and industry structures at the meso-level, and macro-level changes have their own dynamics but also interact. Again, there is unpredictability, which confounds modernistic attempts at intervention to achieve a goal. Yet, understanding of the multi-level dynamics is important and scenarios can be drawn up, as we have shown in the case of nanotechnology (Rip & Te Kulve, 2008).

The Quest to Intervene: Influencing Technological Developments at an Early Stage

In the overall co-evolution of technology and society, a variety of actors are interested in influencing technological change from the perspective of their own goals. Firms will think in terms of market success and strategic advantages. NGOs have their values to pursue; for example, health improvement or a clean environment. National governments and other government agencies have overall goals such as security, quality of life and sustainability, under which a variety of actions are formulated and implemented. Their assessments of the situation, and their actions and interactions, contribute to the co-evolution.

Actors interested in intervening will do better – also in achieving their avowed goals – when they draw on an understanding of the co-evolutionary nature of the overall process. They will then give up on unproductive command-and-control approaches. I note that there are paradoxes: such reflexive action may change the nature of the co-evolutionary processes – so that earlier understanding of the overall process is no longer sufficient. The basic point remains that there is an advantage in understanding the nature of the processes that one is part of and is trying to modulate.

There is a particular challenge, not just for technology developers but increasingly for policy makers and critical societal groups as well: how to influence technological change at an early stage, when irreversibilities have not yet set in and when one can still hope to sway the balance between desirable and undesirable impacts – for oneself (as in the case of the battle for an industry standard, e.g. video recording) or for society more generally.

One problem, clearly, is the difference between functionalities originally envisaged for a technology and the eventually dominant ones that emerge. The branching model, exemplified above in the case of telephony, tells you there will be branching but not which branches will emerge and become dominant. It sensitizes actors to the complexities and so may help to avoid unproductive action. Interestingly, actors learn from past experiences with the dynamics of technological change; for example, in the recent struggle for new DVD standards (Blu-ray vs. HD-DVD) where actors and media refer back to the earlier video-recording standards battle.[1]

Linked to the cognitive problem of not being able to predict the eventual shape of a technology and its context, there are two action-related dilemmas. The so-called anticipation dilemma refers to firms and other technology actors who might prefer to postpone action until the situation is clearer but cannot equivocate indefinitely (Verganti, 1999). For the societal level, where governments, NGOs or societal groups may want to influence technological developments to avoid or limit negative impacts, Collingridge (1980) formulated the knowledge-control dilemma. The dilemma is not unique to technological innovation in context but is readily apparent there: at an early stage, when technological developments are still fluid and amenable to some control, it is unclear what impacts will occur. By the time the first impacts become visible, there are so many irreversibilities, including vested interests, that it is impossible to effect changes for the better.

One can redefine (but not resolve) the dilemmas by recognizing that the dichotomy (especially in the knowledge-control dilemma) is artificial. There are ongoing developments in which assessments are made all the time, tentative steps are taken, and learning from these experiences occurs. As with Lindblom's incrementalist approach to policy (Lindblom and Woodhouse, 1993), there is no guarantee that the path that emerges is a 'good' path. Additional reflexive learning is important (even if it will not provide a guarantee). So-called Constructive Technology Assessment approaches support such reflexive learning, and they address the additional problem of the variety of actors and their possibly contrasting perspectives.[2]

Learning is fractured by the contrast between 'insiders' who invest in developing new technology, and 'outsiders' who receive the results, whether they like them or not. This is not just the general contrast between producers (of a new product) and consumers (with their own, partially articulated preferences). Especially in R&D-based innovations, the development trajectory optimises the new process or product *per se* but its eventual success requires re-contextualisation; a process that cannot

---

1   *Financial Times* (26 August 2005) had an article headed: "Repeat of VHS versus Betamax conflict looms", and noted that "Mr. Nishida [Toshiba's new President] said that the two sides had not given up trying to agree a deal on technical standards, which would avoid a repeat of the video technology battle of the 1970s, when Sony's Betamax lost out to the VHS format."

2   Rip, Misa and Schot (1995) set out the general idea and Schot & Rip (1997) outline generic strategies, including Strategic Niche Management. There is now also 'real-time technology assessment' (Guston and Sarewitz, 2002).

be anticipated fully, let alone determined, at the earlier stages of the trajectory. A striking example is the negative reception by the deaf community of cochlear implants – because this technology undermined their culture (Reuzel, 2004). The example also shows that the problem is not just a cognitive one (how to anticipate the unpredictable) but also a socio-political one (technological development is often a matter of insiders but will ultimately be exposed to outsiders). And it is not just a matter of myopia on the part of inside actors. There is a real dilemma because there are costs involved in taking wider contexts into account at an early stage.

Garud and Ahlstrom (1997), who discuss a number of such examples, draw attention to what they call 'bridging events' between insiders and outsiders. A similar point was made for the role of outsiders and of alternative products in attempts to change existing technical regimes (Van de Poel, 2000). The 'bridging events' might become institutionalized and then create a 'nexus' between the variation and selection components of the co-evolution of technology and society. The concept of 'nexus' was introduced by Van den Belt and Rip (1987) to show how test labs in the synthetic dye industry in the late 19th century were a way to anticipate on the selection environment. After a time, the dynamics were reversed, when dyers started to follow the instructions of the synthetic dye manufacturers. The concept of nexus can be applied more broadly to cover all types of institutionalised couplings between variation and selection.

These are general considerations. The key point to be drawn is that actors who understand co-evolutionary dynamics and emerging patterns can play on them to modulate what happens and perhaps realize some of their goals – a modest modernism. I will show how this point can work out by focusing on a concrete and recurrent pattern: that of an innovation 'journey', as Van de Ven et al. (1999) use the notion. I will mobilize further findings from economics and the sociology of innovation to expand the use of the concept to include contexts and their changes. My analysis will, in a sense, be a demonstration of the possibility of identifying opportunities for intervention, based on an understanding of how the innovation journey is articulated over time, and how it can be mapped anticipatorily because of the existence of recurrent patterns.

The Innovation Journey

Van de Ven and his collaborators introduced the idea of an innovation journey for product and process development in industrial firms (and in networks including R&D institutes) to capture the fact that there are lots of contingencies, shifts and set-backs during the innovation process, and planning and management have only limited effect (Van de Ven et al., 1999). Actual developments were traced retrospectively in the Minnesota Studies on Innovation (Van de Ven, 1989) and used as an argument – as a mirror, as it were – for actors to recognize the complexities. The mapping of the complexity of a journey is necessary because actors tend to project a linear future, defined by their intentions, and use this projection as a road map – only to be corrected by circumstances. Although this is evolution – trial-and-error variation and selection dynamics – one can try to do better.

Conceptually, the innovation journey is a cross-section of the overall co-evolution, and one which has travelled following the enactors of innovation. Given the dominant role of enactors of innovation in determining what will happen, the recognition of overall patterns in such innovation journeys is important for other actors (governmental agencies, NGOs, societal groups) as well, when they want to exert some influence. Figure 1 assembles and integrates results of a large number of studies on industrial product and process innovation. The background of this particular methodology of mapping the innovation journey in context has been described elsewhere (Callon et al., 1992, on techno-economic networks; Rip and Schot, 2002, building on that type of analysis).

Drawing heavily on Rip and Schot (2002), I will tell the 'story' of the innovation journey. It starts with the identification of an opportunity: a new technological option, or the pressure to find a solution to a problem. New options may derive from R&D findings or scientific advance in general but other sources remain important. The role of science varies but has often to do with the discovery or modification, in the laboratory, of an effect that is linked to potential applications (cf. also Robinson et al., 2007). Such discoveries are widely published and pursued in many places without necessarily leading to innovation journeys (Van Lente, 1993). An example is the discovery of high-temperature superconductivity, which led to speculation about more efficient magnetic trains (eventually, other applications of this new laboratory phenomenon proved to be more realistic, e.g. detection systems for very weak magnetic signals). New options can be actively sought, as in the pharmaceutical industry, where the search for 'leads' is a recognized activity. This has to do with the level of articulation of functions to be fulfilled: (re)searchers have a good idea of what they should be looking for. In other sectors, functions – and thus, the search for opportunities – are articulated in a more *ad hoc* manner.

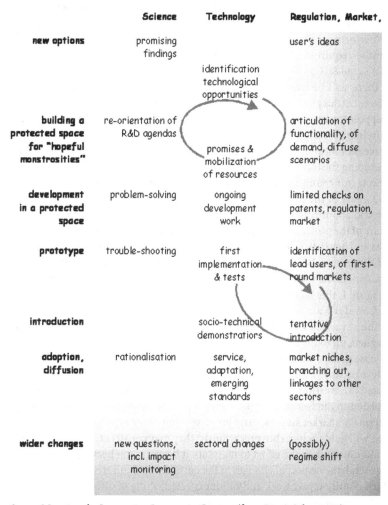

	Science	Technology	Regulation, Market,
new options	promising findings		user's ideas
		identification technological opportunities	
building a protected space for "hopeful monstrosities"	re-orientation of R&D agendas	promises & mobilization of resources	articulation of functionality, of demand, diffuse scenarios
development in a protected space	problem-solving	ongoing development work	limited checks on patents, regulation, market
prototype	trouble-shooting	first implementation & tests	identification of lead users, of first-round markets
introduction		socio-technical demonstrations	tentative introduction
adoption, diffusion	rationalisation	service, adaptation, emerging standards	market niches, branching out, linkages to other sectors
wider changes	new questions, incl. impact monitoring	sectoral changes	(possibly) regime shift

Fig. 1   Mapping the Innovation Journey in Context (from Rip & Schot 2002)

For my analysis, the key point is that such technological opportunities always start out as 'hopeful monstrosities': full of promise but not able to perform very well. Actors will make more specific promises (to internal and external sponsors) to mobilize resources to be able to work on the technological option and nurture it into a semblance of functionality – what is called 'proof of principle'. Such promises anticipate and thus further articulate functions and possible societal demand. Since they also specify what the material, device or artefact should be able to do (its 'performance'), an R&D agenda will be identified and somewhat stabilized. A promise-requirement cycle is launched and shapes the trajectory of development (Van Lente & Rip, 1998).

There will be exaggeration and hand-waving claims, and actors can adopt different strategies: embrace hype or underplay the promise for fear of creating disappointments and a backlash. (This was clear in biotechnology and is occurring again in nanotechnology, cf. Rip, 2006b.) The net effect of the networking and resource mobilisation is the emergence of a protected space for promising R&D and developing the technological opportunity, inside an organisation or across organisations (for example, in the development of a military aircraft, Law and Callon, 1992). Part of the protection stems from a (precarious) agreement over a diffuse scenario about functions to be fulfilled and their envisioned societal usage. The nature of the protected space, together with its boundary agreements, including the rules and heuristics derived from the promises that were made, determine choices and directions. Work within the protected space thus proceeds according to its own dynamics, with only occasional checks with the scenario of usage (if at all). In Figure 1, this is visualized as a cycle, giving rise to temporary path dependency (cf. Garud & Karnoe, 2001).

The advantages of a protected space are clear. Such spaces can be consciously designed, up to the extreme version of a 'skunk works', as when IBM constituted a separate group, with its own resources and outside regular management control, to develop its PC. The risks derive from the fact that the diffuse scenario that legitimated the creation of the protected space in the first instance is not updated. This is related to the fact that, in the early stages, the market/society pole only figures in terms of market studies, early promises, and other expectations. Such expectations will guide actions but precariously: they may construct a 'house of cards' that breaks down when the effort to maintain it becomes too heavy (Van Lente, 1993). In general, the neglect of changes in the outside world can create problems later on, which require repair work and/or (unexpected) shifts in direction (Deuten & Rip, 2000). And, sometimes, the promise will turn out to be empty after all.

The second cycle indicated in Figure 1 also gives rise to some path dependency and thus another trajectory in the journey. At some moment in time, a decision is taken (or emerges) to go for prototypes or other attempts at demonstrating a working technology. Activities include exploring/optimizing production, with trouble shooting and rationalization through further research; implementation of the product and learning about usage, and preliminary market or demand testing. As in the case of biotechnology, regulation and acceptability can become real issues and might direct efforts in particular directions (as happened, for example, in genetic modification of corn and wheat with the choice of herbicide and pesticide resistance rather than of disease resistance).

These activities are much less self-contained than in the earlier research and development trajectory, and subject to more intra- and inter-organizational tensions. Thus, other interactions and management styles are called for. With the tentative introduction of the new product or process, with a few customers ('lead users'), or in a 'societal' experiment (and then often in collaboration with public authorities, as in the case of electric vehicles, cf. Hoogma et al., 2002), the complexity increases but also the opportunities for real-world learning and subsequent modification of the product. Because specific socio-technical couplings are introduced, path dependencies may occur: when dedicated niches are created for learning, the kind of learning depends on the nature of these niches, and this may not be adequate to the demands and selections in the wider world. In addition, as the visibility of the project/the product increases, there may be reactions and repercussions.

Market introduction, an important concern for marketeers and for higher levels in a firm, is thus a process rather than a point decision. The 'market' is neither one-dimensional nor homogeneous, and demand is only gradually articulated in response to supply. There must be something like a protected space (a niche) for the new product, so that it will survive an otherwise too harsh selection, in general and because it might go against the dominant regime.[3] At the same time, when limited to the particular protected space, a product is created that can survive only within that space. This may well be the final outcome: a product for one (and limited) market niche. For a long time, fuel cells, in spite of their recognized general promise, functioned only in space applications. Because of shifts in the motor-car world, a 'branching' occurred and fuel cells are now adopted in earnest to drive motor cars.

By the time branching and wider embedding in society occurs, a third phase of the journey emerges. The dynamics depend on the sector and on wider developments, just as much as on the introduction and diffusion of the innovation

---

3   Every innovation that actually or potentially goes against a dominant regime will have difficulty in de-aligning existing linkages and ways of doing things, even when such alternatives have some credibility, for example, because of a promise of sustainability. Re-aligning requires room for experimentation and learning, which is not freely given. The general approach of Strategic Niche Management can be used (Kemp et al., 2001), even if one has to be careful and avoid a simplistic David *versus* Goliath storyline: the tiny novelty conquering the mighty regime (Stuiver 2008). As Hoogma (2000) has shown in the case of electric vehicles, a partial 'fit' strategy is often the only possibility at an early stage and, only after some niches have been created, can the more ambitious 'stretch' strategy be successful (if at all).

itself. There is a cumulative effect of further varieties of application, of suppliers orienting themselves to the new technology, of economies of scale and scope that are exploited, and of recognition by users of further possibilities, which creates new socio-technical linkages. The sector starts to change and its relations with other sectors change. The latter can become so important that the technology driving such changes is recognized as a pervasive technology and characteristic of a new techno-economic paradigm (Freeman & Soete, 1997).

Cumulative effects may lead to the emergence of new regimes and/or shifts in existing regimes. This is a multi-actor, multi-level process, in which no single actor can sway the balance intentionally. Of course, actors will attempt to do so; they will jockey for position in the newly emerging games and regimes, and involve themselves in strategic alliances. In standard setting in information and communication technology and in consumer electronics, such processes are very visible. These then become part of broader socio-technical regime changes. One can argue that a third 'cycle' can result from regime change, when further developments are shaped by the 'grammar' of the new regime. In fact, a number of authors have analysed historical cases in these terms (Van den Ende & Kemp (1999) on the computer regime, and Geels (2005) on steamships, motorcars, and jet engines). Regime change and its stabilization can be presented as a victory. Indeed, the actors involved, as well as the media reporting of the struggles, may think in terms of heroic stories in which power and cleverness of (single) actors determine the outcome. While there definitely is a role for 'regime entrepreneurs', the cumulative process of increasing interdependencies and sunk investments is decisive.[4]

In this stylized description of the innovation journey, three clusters of activities can be distinguished where the innovation journey enters a new phase, and a trajectory with its own dynamics is launched:

- build-up of a protected space,
- stepping out into the wider world,
- sector-level changes.

Each phase has its own dynamics and the trajectories are not easy to modify. But, before the path 'gells', it is easier to exert influence while there is also some assurance that a real difference will result because the intended shift becomes part of the trajectory. In other words, attempts to influence the innovation journey – i.e. to help shape its further evolution – should focus on the entering of a new phase as the *locus* where a

---

4   Compare how Garud et al. (2007) speak of (socially) embedded entrepreneurs.

difference can be made (Rip & Schot, 2002). This is a way to address the dilemmas of Verganti and Collingridge with which this section opened.

Discussion: Intelligent Intervention in Innovation Journeys

Making a difference is not a matter of brute force but of intelligent intervention – i.e. playing with the dynamics and drawing on a sophisticated understanding of them. This applies to attempts at intervention by insiders as well as outsiders. Modulation (with some orchestration) of the dynamics appears to be the right approach. With the many actors involved, and the heterogeneity of their interests and strategies, there is no guarantee that a coherent direction will evolve. Also, waiting games may emerge, which create impasses even when there is a sense of direction.

A 'shadow of authority' may be necessary to break through impasses. Credibility pressures, for example in relation to environmentally-friendly products and processes, can also play such a role. Authority and credibility pressure are also routes through which public interest considerations about desirable directions can be brought to bear on the dynamics of development. While this indeed happens, for example, in 'technology forcing' regulation by governments and in public debates and consensus conferences about new technologies and their eventual impacts, there are also limitations because of the outsider position.

Governmental and societal actors face the further problem of evaluating the dynamics and possible directions of an innovation journey (often, a cluster of innovation journeys – say, in the genetic modification of plants) from the outside and relating them to their own (often not completely articulated) views of what is desirable. Such views can relate to wealth creation, policy learning and/or risk avoidance, and there may well be conflicts between the various views and goals. The key point is that not only the eventual performance, uptake and use of the new technology and products but also the views and preferences related to the new possibilities are unknown at first. In other words, Collingridge's knowledge and control dilemma is actually a trilemma, with the third horn related to uncertainty about desirable directions. The third horn can be turned into an opportunity for policy articulation and engagement with civil society, as has actually been proposed by a European Union Expert Group looking at the challenges of converging technologies (Nordmann et al., 2004). The strategy articulation workshops that we have designed and organized, focusing on domains of nanotechnology and their contexts, are also an occasion – in the small – of addressing the third horn of the

Collingridge trilemma (Van Merkerk & Robinson. 2006; Robinson & Propp, 2008; Rip & Te Kulve, 2008).

Here too, understanding the dynamics of developments is necessary to arrive at adequate insights and realistic action. This leads to the further question of the robustness of the co-evolutionary perspective and, in particular, the approach employing innovation journeys, as laid out in the previous section. There are two basic points that must be addressed: how to turn essentially retrospective insights concerning what happens in innovation journeys into prospective suggestions for action. And how generalizable is the pattern of the innovation journey as summarized in Figure 1.

First, the mapping of innovation journeys, as carried out by Van de Ven et al. and in our own case studies of nanotechnology domains, and also in the many historical, sociological and economic case studies of innovation, is essentially retrospective. It shows what we know about such processes as they have occurred. In the beginning, however, only the items at the top of the map (Figure 1) are in place, and the future is uncertain. Over time, when the innovation journey progresses, further parts of the pattern shown on the map materialize. At an early stage, actors will rely on their own projections of how innovation processes usually go, and what is necessary to make them successful, to guide them in their actions. Things might still go wrong, of course, and for a variety of reasons. One cluster of reasons is the limitations of actors' views: because of their limited experience, because of a tendency to be myopic (which can be reinforced by the context, cf. Pavitt, 1990), and because of neglect of what is already known. Thus, having an overall map of the general pattern in the innovation journey (if and when relevant to their situation) will help them to do better.

This 'enlightenment' model applies to all actors: technology developers or 'enactors' (Garud & Ahlstrom, 1997), policy makers and other government actors, 'third parties' such as venture capitalists and insurance companies, and civil society actors. We have been involved in developing mapping tools to support 'enactors' of innovation, which introduce complexity (in particular, by forcing actors to specify the context of the innovation and taking it into account) and thus enable them to do better.[5] The key element that justifies a knowledge-and-understanding-based prospective approach (in contrast to brainstorming about the future) is the notion of 'endogenous futures' (Rip et al., 2007, Rip & Te Kulve, 2008). Co-evolutionary

---

5   The tools range from translating insights from economics and the sociology of technology (Rip 1995), and visualizations of concentric biases (Deuten et al. 1997), to more detailed project management support in the SocRobust project, where we offered ways to draw managers out of their concentric bias (Larédo et al. 2002). Our recent scenario exercises (Rip and Te Kulve 2008) are a further step.

dynamics do not stop in the present where actors consider their actions as if they were free agents. In other words, futures are predicated on what has been happening. Our understanding of such co-evolutionary dynamics is sufficiently advanced to articulate what such futures might be. In the case of path dependencies, leading to trajectories, this point is widely recognized. Broadening the notion of path dependence to 'entanglements' (Rip et al., 2007) allows further application of the general idea, even if the prospective mapping should not be linear (that is a reason to work with scenarios). The innovation journey pattern, as discussed in the previous section, can now be seen as a way of recognizing, at an early stage, endogenous futures.

Second, the question of generalizability of the pattern. My analysis of possibilities for modulating an innovation journey was carried out for a quite specific context: that of technological developments initiated in firms or other technology-promoting organizations. In other contexts, we might find other dynamics. This need not undermine the general suggestion of intelligent intervention focused on specific *loci* but their specification, and the shape of the intervention, will have to be different.

Further, I am prepared to claim that there are only a limited number of different patterns in innovation journeys. My argument for this claim is contextual. The reason the innovation journey for industrial product and process innovations has a recognizable and recurrent pattern is the way industrial innovation has become institutionalized in our societies. The pattern is reproduced because it is seen as the natural way to do things and because it remains productive. Other patterns that might occur will also depend on the nature of the innovations and the nature and extent of institutionalization, and that is how we can identify them.

I will discuss innovation in agriculture, taken to include nature and rural development. The general features of innovation, such as the introduction of novelty and the de-alignment and re-alignment that occurs and has to occur, will be visible there as well. A specific feature is how natural processes in the soil, in plants and in animals and their interactions, are an integral part of the configuration that has to work. 'Natural history' is more important than mechanical and electrical engineering, and design and construction take on a different complexion. Natural processes are variable and locally contingent, so that what works on a particular location and at a particular time may not work elsewhere and at another time. And one has to wait and see what grows; one cannot simply force one's intended re-alignments on nature.

Thus, while protected spaces for further development such as field experiments are possible, they will not always deliver results that are transferable to other fields – unless these can be transformed so as to be identical to the original field of the experiment. While nature is always a partner in the co-production of innovation and its impacts, there are different ways of 'managing' the partnership. One route is based on experiments in laboratories and agricultural stations (so the protected spaces are more or less controlled), and novel options need to be transferred, precariously, to local practices. This is mostly done by colonizing these practices and transforming them into something similar to the experimental plots (the traditional route of agricultural research and extension since the late 19th century). There is increasing interest in another route, that of modifying the new option rather than modifying the local practice, because of the failures that occurred in the later route (Wiskerke & Van der Ploeg, 2004). Recognizing this, one can start at the other side with novelties generated in local practices and consider that as the basic pattern. The currently dominant expert-based regime would then be seen as an overlay on the basic pattern (Van der Ploeg, 2003), and its failures – for example, the grinding down of the Green Revolution – would be seen (and explained) as structural because of the neglect of the basic 'natural history' pattern (and of social dynamics).

The basic pattern is still alive, even if it has to survive in a modernist world. Local novelties will confront a dominant regime not only in terms of how to justify and develop themselves but also in relation to rules and regulations derived from the dominant regime. An example is how new ventures in low-input agriculture, including new ways of handling manure (in Friesland, one of the Dutch provinces), had to struggle against regulations specifying how ammonia emissions should be reduced. It took network building and concerted action to gain exemption from the regulation and demonstrate the promise of the new approach (Stuiver, 2008). A protected space was created for further experimentation but it was geared to the contingencies of local situations and their history rather than controlled construction and performance measurements, as in the industrial product development pattern. As case studies in South Africa have shown, novelties need to be protected against the traditions and exigencies of the local situation to allow them to grow – and an additional ambivalence is introduced by the role of sponsors and donors creating such a protected space, which also implies dependency relationships that cannot always be overcome (Adey, 2007).

Modern agriculture (including nature and rural development) is in flux, and patterns that were common before the Western expert-based regime asserted itself, in the West and then elsewhere, are accepted again. This is also visible in the debates on (and contestation of) high-tech inputs in agriculture, where participatory approaches and integration of local knowledge are increasingly welcomed.

If one looks at innovation in agricultural machinery and agro-chemicals, and now biotechnology, one will see innovation journeys resembling the pattern of industrial product development. However, if one starts with local practices – or just with the difficulties of handling natural variability and contingency to serve human purposes – another pattern is visible, where innovation on location, including social innovation, is the driver. The expert-based and high-tech routes are an addition to this pattern. Actually, when one looks closely, there are many interesting cases where the ultimately successful innovation consists of local practices developing endogenous technology together with selective use of imported technology (Bertelsen & Müller, 2003).

It is possible to identify and explore patterns of innovation journeys in other domains. Concretely (and realizing that I am reducing actual complexity), I suggest there are four main types of innovation journey, their differences depending on the institutionalisation of innovation in various sectors and their historically evolved constellation of actors, and on the nature of the relevant technologies-in-context. One reason why constellations of actors are important is their role in creating protected spaces.

The four basic types of innovation journey are:

1. industrial product and process innovation, where firms recognize that dedicated effort is necessary rather than just learning from practices.
2. energy, transport and infrastructure, where long lead times are taken into account, and there is little or no possibility of testing the overall system in a protected space. The real world becomes the laboratory (cf. Krohn & Weyer, 1994).
3. ICT, with its dual structure (carriers and services), big and small players, and the key role of software engineering and orgware and socioware aspects.
4. 'natural history' as exemplified in agriculture (and nature conservation, environment, and rural development) but also in medical practices.

There are overlaps and mixed dynamics, as was clear in the case of agriculture. Transitions towards, for example, a hydrogen economy follow Pattern 2 but are enabled by Pattern 1 innovation journeys as in fuel cells and hydrogen storage devices. The medical and health domain is such a mixed-dynamics case, I would claim, so that it would not qualify as a further basic type of innovation journey. Earlier practice-based innovation patterns in this domain were backgrounded by suppliers of apparatus and drugs, even when the role of professional users remained important (Blume, 1992). While Pattern 1 may now be dominant, the increasing importance of health care is the reason that the practice orientation is coming back. The overall pattern may well become like the one for agriculture – handling variability and local contingency of natural processes, now of humans rather than soil, plants and animals.

Of course, each of the types of innovation journey can (and should) be described in as much detail as I provided for the first industrial product and process innovation, and as I started out to do for agriculture. For intervention, the key point is to forget about one-size-fits-all approaches and recognize that innovation patterns can and should be disaggregated to be able to address ('manage') them productively. Pavitt (1984)'s early analysis of 'innovation patterns' has the same thrust but he refers to largely stabilized industry structures rather than to the dynamics of innovation processes.[6]

Clearly, there are recurrent patterns in the co-evolution of innovation, institutions and society. Going for modulation rather than command-and-control, there are still varieties of modulation. Taking one pattern as the entrance point, the innovation journey in the context of industrial product development allowed me to identify interesting possibilities for modulation: *loci* for making a difference, protected spaces and strategic opening of such spaces. Additional research and further reflection is necessary, especially to discover if the overall approach of looking at dynamics and ways to modulate the dynamics can be productively applied to energy/infrastructure and ICT.

Coda

I have come a long way since 1977 when the Nelson & Winter article was published and some governments adopted their message in their technology and innovation policy. In a sense, innovation has come a long way as well. To paraphrase the title of a chapter in Felt and Wynne (2007), innovation has been reinventing itself including, on the one hand, open innovation around promising high tech (cf. journey type 1) and, on the other hand, 'collective experimentation' as in open-source software (cf. journey type 3), and the involvement of patient associations and farmers' collectives (cf. journey type 4). Other authors like Von Hippel (2005) and Malerba (2006) have indicated changes in innovation patterns (for innovation journeys type 1 and type 3). For the type 2 innovation journey, long-term planning and coordination, now with public-private alliances (and a neo-corporatist governance pattern), continues.

Another important change is the interest in anticipatory coordination in the face of uncertainties about new developments. The advent of European Technology Platforms for particular areas is a strong indicator. While they are established for all

---

6   Van de Poel (1998) introduced a (necessary) modification to Pavitt's typology by including a mission-driven pattern where the role of government is important. There is also the extension of the typology to services, as in the often quoted paper by Soete and Miozzo (1989). Unfortunately, within the scope of this chapter, I cannot address innovation in services explicitly.

sorts of domains, there is an emphasis on capturing promising high-tech domains such as nanotechnology, embedded systems and the hydrogen economy (and their innovation journey type 1 dynamics). Such anticipatory coordination adds to the reflexivity of co-evolution. When there are effects, it will be through modulation, and this then becomes an integral part of 'doing' innovation. It is a further example of institutionalized coupling between variation and selection, i.e. a nexus.

So what is the role of government? The basic point that government is only one of the actors in the overall evolution of multi-actor, multi-level systems and their interactions is increasingly recognized. Government does have a special responsibility: for infrastructures and other collective goods, and for the conditions that allow for productive and desirable innovations. Implementing that responsibility through generic measures, as neo-classical economy would suggest, runs up against actual variety. Specific measures will, of course, suffer from information asymmetries. The problem (a dilemma?) can never be resolved completely but the reduction of complexity that I have proposed, of thinking and acting in terms of a limited number of patterns of innovation journeys, will go some way to addressing that problem.

The traditional justification for government technology and innovation policy was market failure. There is government failure as well. In a co-evolutionary perspective, the important issue is selection-environment failure and the need to address such failures at the system level (Smits & Kuhlmann, 2004), where government has a role to play anyway. Stimulating nexuses and some monitoring of their productivity would be a new approach; or better, it occurred already but is now recognized for what it can do and why. An example is the establishment of Technology Assessment organizations as a link to selection environments through anticipation and interaction.[7] The recurrent discussions of patenting and intellectual property rights could also usefully be seen as being about the productivity of nexuses. Stimulating horizontal interactions between different kinds of actors, for a purpose, could also be designed and evaluated in terms of co-evolution, including the way government agencies are increasingly part of the consortia and forums that are formed.

These brief considerations should be located in the general thrust of my argument about understanding and intervention. There are patterns and dynamics to base policies and actions on, even if these patterns are contingent in the sense that they are outcomes of historical co-evolutionary processes. Modulation is what intelligent intervention should be about, by government actors or any other actor.

---

7   Freeman and Soete (1997), in their conclusion, put up a similar argument and suggest that government technology policy should limit itself to Constructive TA.

Bibliography

Abernathy, W. J., & Clark, K. B. (1985). Innovation: Mapping the winds of creative destruction. *Research Policy, 14*, 3-22.

Adey, S. (2007). *A journey without maps: Towards sustainable subsistence agriculture in South Africa*. Wageningen: publisher not identified.

Basalla, G. (1988). *The Evolution of Technology*. Cambridge: Cambridge University Press.

Bertelsen, P., & Muller, J. (2003) Explicating the indigenous systems of innovation in Tanzania. In M. Muchie, P. Gammeltoft & B.-A. Lundvall (Eds.), *Putting Africa first. The making of African innovation systems* (pp. 123-138). Aalborg: Aalborg University Press.

Blume, S. S. (1992). *Insight and industry. On the dynamics of technological change in medicine*. Cambridge: Cambridge University Press.

Borup, M., Brown, N., Konrad, K., & van Lente, H. (2006). The sociology of expectations in science and technology. *Technology Analysis & Strategic Management, 18*(3/4), 285-298.

Brown, N., Rappert, B., & Webster, A. (Eds.). (2000). *Contested futures. A sociology of prospective techno-science*. Aldershot etc: Ashgate.

Callon, M., Laredo, P., & Rabeharisoa, V. (1992). The management and evaluation of technological programs and the dynamics of techno-economic networks: The case of the AFME. *Research Policy, 21*, 215-236.

Callon, Michel (Ed.). (1998). *The laws of the market*. Oxford: Blackwell.

Callon, M. (2002). From science as an economic activity to socioeconomics of scientific research. The dynamics of emergent and consolidated techno-economic networks. In P. Mirowski & E.-M. Sent (Eds.), *Science bought and sold. Essays in the economics of science* (pp. 277-317). Chicago and London: University of Chicago Press.

Collingridge, D. (1980). *The social control of technology*. London: Frances Pinter.

Cusumano, M. A., Mylonadis, Y., & Rosenbloom, R. S. (1992). Strategic maneuvering and mass-market dynamics: The triumph of VHS over Beta. *Business History Review, 66*(01), 51–94. https://doi.org/10.2307/3117053

Deuten, J. J., Rip, A., & Jelsma, J. (1997). Societal embedment and product creation management. *Technology Analysis & Strategic Management, 9*(2), 219-236.

Deuten, J. J., & Rip, A. (2000). Narrative infrastructure in product creation processes. *Organization, 7*(1), 67-91.

Dosi, G. (1982). Technological paradigms and technological trajectories: A suggested interpretation of the determinants and directions of technical change. *Research Policy, 11*, 147-162.

Felt, U., Europäische Kommission, & Europäische Kommission (Eds.). (2007). *Taking European knowledge society seriously: report of the Expert Group on Science and Governance to the Science, Economy and Society Directorate, Directorate-General for Research, European Commission*. Luxembourg: Off. for Official Publ. of the Europ. Communities.

Freeman, C., & Louca, F. (2001). *As time goes by. From the industrial revolutions to the information revolution*. Oxford: Oxford University Press.

Freeman, C., & Soete, L. (1997). *The economics of industrial innovation* (3rd ed.). Cambridge, MA: MIT Press.

Garud, R. (1994). Cooperative and competitive behaviors during the process of creative destruction. *Research Policy, 23*, 385-394.

Garud, R., & Ahlstrom, D. (1997). Technology assessment: A socio-cognitive perspective. *Journal of Engineering and Technology Management, 14*, 25-48.

Garud, R., & Karnoe, P. (2001). *Path dependence and creation.* Mahwah, N.J.: Lawrence Erlbaum Associates.

Garud, R., Hardy, C., & Maguire, S. (2007). Institutional entrepreneurship as embedded agency: An introduction to the special issue. *Organization Studies, 28*(7), 957-969.

Geels, F. W. (2002). *Understanding the dynamics of technological transitions: a co-evolutionary and socio-technical analysis.* Enschede: Twente University Press.

Geels, F. W. (2005). *Technological transitions and system innovations. A co-evolutionary and socio-technical analysis.* Cheltenham: Edward Elgar.

Guston, D., & Sarewitz, D. (2002). Real-time technology assessment. *Technology in Culture, 24*, 93-109.

Hoogma, R. (2000). *Exploiting technological niches: strategies for experimental introduction of electric vehicles.* Enschede: Twente University Press.

Hoogma, R., Kemp, R., Schot, J., & Truffer, B. (2002). *Experimenting for sustainable transport. The approach of strategic niche management.* London: Spon Press.

Kemp, R., Rip, A., & Schot, J. (2001). Constructing transition paths through the management of niches. In R. Garud & P. Karnoe (Eds.), *Path dependence and creation* (pp. 269-299). Mahwah, N.J.: Lawrence Erlbaum Associates.

Krohn, W., & Weyer, J. (1994). Society as a laboratory: The social risks of experimental research. *Science and Public Policy, 21*(3), 173-183.

Laredo, P., Rip, A., Jolivet, E., Shove, E., Raman, S., Moors, E. H. M., ... Garcia, C. E. (2002). *SocRobust (Management tools and a management framework for assessing the potetial long-term S&T options to become embedded in society) Final Report; Project SOE 1981126 of the TSER Programme of the European Commission.* Paris: Armines.

Law, J., & Callon, M. (1992). The life and death of an aircraft: A network analysis of technical change. In W. Bijker & J. Law (Eds.), *Shaping technology / building society* (pp. 21-52). Cambridge, MA: MIT Press.

Lente, H. van. (1993). *Promising technology: the dynamics of expectations in technological developments.* Delft: Eburon Publ.

Lente, H. van., & Rip, A. (1998). Expectations in technological developments: An example of prospective structures to be filled in by agency. In: C. Disco & B. J. R. van der Meulen (Eds.), *Getting new technologies together* (pp. 195-220). Berlin, New York: Walter de Gruyter.

Levinthal, D.A., 'The Slow Pace of Rapid Technological Change: Gradualism and Punctuation in Technological Change', *Industrial and Corporate Change* (1998) 217-247.

Lindblom, C. E., & Woodhouse, E. J. (1993). *The policy-making process* (3rd ed.). Englewood Cliffs, NJ: Prentice Hall.

Malerba, F. (2006), Innovation and the Evolution of Industries, *Journal of Evolutionary Economics* 16 3-23.

Mokyr, J. (1990). *The lever of riches: Technological creativity and economic progress.* Oxford University Press, Gary, NC. 349 pages. ISBN: 0-19-606113-6. $NA. (1992). *Bulletin of Science, Technology & Society, 12*(3), 183–183. https://doi.org/10.1177/027046769201200318

Nelson, R. R., & Winter, S. G. (1977). In search of a useful theory of innovation. *Research Policy, 6*, 36-76.

Nelson, R. R., & Winter, S. G. (1982). *An evolutionary theory of economic change.* Cambridge, Mass.: Belknap Press.

Nelson, R. R. (1994). The co-evolution of technology, industrial structure, and supporting institutions. *Industrial and Corporate Change, 3*, 47-63.

Nordmann, A., & Europäische Kommission (Eds.). (2004). *Converging technologies: shaping the future of European societies; Report 2004*. Luxembourg: Office for Official Publ. of the Europ. Communities.

Pavitt, Keith, 'Sectoral Patterns of Technical Change: Towards a Taxonomy and a Theory', *Research Policy* 13 (1984) 343-373.

Pavitt, K. (1990). The international pattern and determinants of technological activities. In S. E. Cozzens, P. Healey, A. Rip & J. Ziman (Eds.), *The research system in transition* (pp. 89-101). Dordrecht: Kluwer Academic.

Ploeg, J. D. van der. (2003). *The virtual farmer: past, present, and future of the Dutch peasantry*. Assen: Royal van Gorcum.

Poel, I. van de. (1998). *Changing technologies: A comparative study of eight processes of transformation of technological regimes*. Enschede: Twente University Press.

Poel, I. van de. (2000). On the role of outsiders in technical development. *Technology Analysis & Strategic Management, 12*(3), 383-397.

Pressman, J. L., & Wildavsky, A. (1984). *Implementation* (3rd Edition, Expanded). *How great expectations in Washington are dashed in Oakland*. Berkeley, Cal.: University of California Press.

Reuzel, R. (2004). Interactive technology assessment of paediatric cochlear implantation. *Poiesis & Praxis, 2*, 119-137.

Rip, A. (1995). Introduction of new technology: Making use of recent insights from sociology and economics of technology. *Technology Analysis & Strategic Management, 7*(4), 417-431.

Rip, A., & Kemp, R. (1998). Technological change. In S. Rayner & E. L. Malone (Eds.), *Human choice and climate change* (2nd ed.). (pp. 327-399). Columbus, Ohio: Battelle Press.

Rip, A. (1998) Modern and post-modern science policy. *EASST Review, 17*(3), 13-16.

Rip, A., & Schot, J. W. (2002). Identifying *loci* for influencing the dynamics of technological development. In R. Williams & K. Sorensen (Eds.), *Shaping technology, guiding policy* (pp. 158-176). Cheltenham: Edward Elgar.

Rip, A. (2006a). A co-evolutionary approach to reflexive governance – and its ironies. In J.-P. Vos, D. Bauknecht & R. Kemp (Eds.), *Reflexive governance for sustainable development* (pp. 82-100). Northampton, MA: Edward Elgar.

Rip, A. (2006b). Folk theories of nanotechnologists. *Science as Culture, 15*(4), 349-365.

Rip A., Robinson D. K. R., & te Kulve, H. (2007). *Multi-level emergence and stabilisation of paths of nanotechnology in different industries/sectors. Paths of developing complex technologies: Insights from different industries*. Berlin, September 17-18. Sponsored by the Volkswagen Foundation.

Rip, A., & te Kulve, H. (2008). Constructive technology assessment and sociotechnical scenarios. In E. Fisher, C. Selin & J. M. Wetmore (Eds.), *The yearbook of nanotechnology in society, Volume I: Presenting futures* (pp. 49-70). Berlin etc: Springer.

Robinson, D. K. R., Rip, A., & Mangematin, V. (2007). Technological agglomeration and the emergence of clusters and networks in nanotechnology. *Research Policy, 36*, 871-879.

Robinson, D. K. R., & Propp, T. (2008). Multi-path roadmapping as a tool for reflexive alignment in emerging S&T. *Technology Forecasting and Social Change* (Article in Press March 2008).

Schaeffer, G. J. (1998). *Fuel cells for the future. A contribution to technology forecasting from a technology dynamics perspective*. Enschede: Twente University Press.

Smits, R., & Kuhlmann, S. (2004). The rise of systemic instruments in innovation policy. *International Journal of Foresight and Innovation Policy, 1*, 4-32.

Soete, L., Miozzo, M. (1989). *Trade and development in services: a technological perspective.* Working Paper No. 89–031. Maastricht: MERIT.

Stoelhorst, J.-W. (1997). *In search of a dynamic theory of the firm: an evolutionary perspective on competition under conditions of technological change with an application to the semiconductor industry.* [University of Twente], Twente.

Stuiver, M. (2008). *Regime change and storylines. A sociological analysis of manure practices in contemporary Dutch dairy farming.* Wageningen: publisher not identified.

Van den Belt, H., & Rip, A. (1987). The Nelson-Winter/Dosi model and synthetic dye chemistry. In W. E. Bijker, T. P. Hughes & T. J. Pinch (Eds.), *The social construction of technological systems. New directions in the sociology and history of technology* (pp. 135-158). Cambridge, MA: MIT Press.

Van den Ende, J., & Kemp, R. (1999). Technological transformations in history: how the computer regime grew out of existing computing regimes. *Research Policy, 28*(8), 833–851. https://doi.org/10.1016/S0048-7333(99)00027-X

Van de Ven, A. H., Polley, D. E., Garud, R., & Venkataraman, S. (1999). *The innovation journey.* New York & Oxford: Oxford University Press.

Van de Ven, A. H., Angle, H. L., & Poole, M. S. (Eds.). (2000). *Research on the management of innovation: the Minnesota studies.* Oxford; New York: Oxford University Press.

Van Merkerk, R. O., & Robinson, D. K. R. (2006). The interaction between expectations, networks and emerging paths: A framework and an application to lab-on-a-chip technology for medical and pharmaceutical applications. *Technology Analysis and Strategic Management, 18*(3-4), 411-428(18).

Verganti, R. (1999). Planned flexibility: Linking anticipation and reaction in product development projects. *J. Product Innovation Management, 16*, 363-376.

Visscher, K., & Rip, A. (2003). Coping with chaos in change processes. *Creativity and Innovation Management, 12*(2), 121-128.

Von Hippel, E. (2005). *Democratizing innovation.* Cambridge, MA: MIT Press.

Wiskerke, J. S. C., & van der Ploeg, J. D. (Eds.). (2004). *Seeds of transition. Essays on novelty production, niches and regimes in agriculture.* Assen: Royal Van Gorcum.

Chapter 4
De facto Governance of Nanotechnologies*

New and emerging technologies, especially nanotechnologies with the structural uncertainties about their eventual functionalities and risks, are a challenge to governance. Regulatory agencies in Europe and the USA review existing regulation and consider voluntary reporting as a transitional measure. Risk governance is opened up to include public dialogues and deliberative processes. What is striking is how much actual governance is already occurring in and around nanotechnology without any particular actor being responsible for the emerging governance arrangements.

Thus, the first aim of this paper is to map what is happening: the actions and interactions and how these add up to outcomes at the collective level which function as governance arrangements. In that way, the paper is explorative, it is an attempt to understand what is occurring, and is partly based on the author "moving about" as a self-styled anthropologist in the world of nanotechnologies. What is clear is that the emerging governance arrangements have a distributed character. This is captured by using the notion of governance, which in contrast to government, is distributed almost by definition. The additional point, however, is that bottom-up actions, strategies and interactions are constitutive for these arrangements, rather than that they then the result of opening up an earlier centralized arrangement and make it more distributed – a common way to introduce the notion of governance (Van Kersbergen and Van Waarden 2004). To emphasize the strong bottom-up character of what is happening I introduce the notion of *de facto* governance.

This leads to the second aim of this paper: to flesh out the notion of *de facto* governance by showing how it works in the domain of nanotechnologies. The recognition of the importance of *de facto* governance implies that attempts at regulation can be located as interventions in emerging *de facto* governance, and will depend on

---

* Source: Morag Goodwin, Bert-Jaap Koops and Ronald Leenes (eds.), Dimensions of Technology Regulation, Nijmegen: Wolf Legal Publishers, 2010, pp. 285-308.

© Springer Fachmedien Wiesbaden GmbH, part of Springer Nature 2018
A. Rip, *Futures of Science and Technology in Society*, Technikzukünfte, Wissenschaft und Gesellschaft / Futures of Technology, Science and Society, https://doi.org/10.1007/978-3-658-21754-9_5

it for their effectiveness.[1] This is similar to how Henry Mintzberg (1994) discussed intentional (and often top-down) strategy in firms and other organizations and argued that its effects will depend on the interaction with *de facto*, or in his words 'pattern' and 'emergent' strategies that are out there already. While society should not be seen as an organization, writ large, the dual dynamics outlined by Mintzberg occur all the time. And they can add-up to what one could call a societal agenda.

There is a third aim, linked to what I see as an intriguing potential *de facto* governance pattern: the internalization of requirements of "responsible development" of nanosciences and nanotechnologies in the actual technological and product-development choices and strategies. Something of the sort is happening, as I will show, and the question then is what this implies for the governance of nanotechnologies. The further question is whether internalization of such considerations might occur for other emerging technologies as well. If this occurs, or is expected to occur, it will create a new regime of shaping technology development in society.

After fleshing out the notion of *de facto* governance, I will present and discuss two recent developments. First, how a socio-technical agenda about promises and concerns about nanotechnology emerged, in which risks, and in particular risks of nano-particles, became dominant. And second, how "responsible development" has become an integral part of the discourse, and to some extent, of the practice, of nanotechnology. This then allows me to inquire into the possible internalization of societal considerations in ongoing development of nanosciences and nanotechnologies. In the concluding comments, I will come back to governance issues.

The Notion of De facto Governance

In the broadest sense of the concept of governance, all structuring of action and interaction that has some authority and/or legitimacy counts as governance. Authors like Van Kersbergen and Van Waarden (2004) and Kooiman (2003) recognize this, even if they do not thematize it. Governance arrangements may be designed to serve a purpose, but can also emerge and become forceful when institutionalized. The same move is visible in Voß et al. (2006: 8) when they say that governance refers

---

1   This point has been made in implementation studies, starting with Presmann and Wildavsky (1984) and becoming almost a movement (of the "bottom-uppers") in the 1980s (see Hanf and Toonen 1985). There is a tendency, however, to be background this in actual policy making and implementation – because policy makers must show that it is they who are making a difference.

to "the characteristic processes by which society defines and handles its problems. In this general sense, governance is about the self-steering of society." They then develop this further:

> governance is understood as the result of interaction of many actors who have their own particular problems, define goals and follow strategies to achieve them. Governance therefore also involves conflicting interests and struggle for dominance. From these interactions, however, certain patterns emerge, including national policy styles, regulatory arrangements, forms of organisational management and the structures of sectoral networks. These patterns display the specific ways in which social entities are governed. They comprise processes by which collective processes are defined and analysed, processes by which goals and assessments of solutions are formulated and processes in which action strategies are coordinated. (...) As such, governance takes place in coupled and overlapping arenas of interaction: in research and science, public discourse, companies, policy making and other venues.

This view has been offered before, notably by Elinor Ostrom. As Scharpf (1997: 204) phrases it: "much effective policy is produced not in the standard constitutional mode of hierarchical state power, legitimated by majoritarian accountability, but rather in associations and through collective negotiations with or among organizations that are formally part of the self-organization of civil society rather than of the policy-making system of the state (Ostrom 1990)." A specific aspect is highlighted by Braithwaite and Drahos (2000: 10), when they say: "The global perspective on regulation we promote not only reframes individuals as subjects and objects of regulation (as in the drug case) and states as subject and object of regulation (by Moody's, the IMF, the Rothschilds and Greenpeace). Understanding modernity, we find, demands the study of plural webs of many kinds of actors which regulate while being regulated themselves."[2]

In such an encompassing view of governance, explicit attempts at steering and intentional government arrangements, will be seen as part and parcel of the overall process, not outside it. In economics, one can speak of endogenizing a factor (like new technology) that had been considered as external to economic analysis. Similarly, one can now say that government and design of governance arrangements must be endogenized to capture what is happening (Rip 2006, cf. also Voß 2007).

When governance of technology is discussed, it appears to be reduced to either innovation stimulation or regulation of actual and possible side-effects. The focus on performance of technologies (positive and/or negative) seems obvious, but it backgrounds broader considerations of governance. Science and Technology Studies (STS) have offered case studies and analysis showing that there is actually

---

2   I quote this after Voß (2007: 34).

a lot of broader governance going on, but it requires a lot of work to overcome this prevalent view of it being technically driven and/or naturalized. Actually, since it is a prevalent view, the simple distinction between innovation stimulation and regulation is itself an example of a governance pattern in the broad sense.

This governance patter derives from the gap between development and promotion of technology, and the responses of society which emerged in the industrial revolution and stabilized in the 19th and early 20th century (Rip et al. 1995). A version of the gap is visible in the institutional separation between promotion and control of new science and technology, for example the difference in outlook and activities between government departments for trade and industry on the one hand, and for social, health and environmental affairs on the other hand. To some extent this is a productive division of labour. But the separation of technology development (in firms, in public research institutes, in technical universities) from wider society implies that society has to respond, somehow, and is at a disadvantage because there have been investments in development already.[3] By now, there is recognition of there being these two separate worlds, of "enactors" of new technology vs. (civil) society. And also recognition that with regard to new technology, civil society is "forced" into one or another of three reactions: to welcome it (this appears to occur for large parts of information and communication technology), to be fatalistic (for example about new infrastructural technologies), or to oppose (as happened with agricultural biotechnology). The recent interest of technology "enactors" in engaging civil society and having public dialogues on new technologies can be seen as an attempt to improve the chance that society will respond in the welcoming rather than the oppositional mode.

Another example of *de facto* governance arrangements, and one which has been highlighted by STS studies, are sociotechnical systems and infrastructures which add up to the sociotechnical landscapes in which we live and move about. Roads and motorways serving automobile transportation, and the structures linked to them, are a clear example of how these "arrangements" shape what we do and cannot do, and with the authority that comes with their being invisible because self-evidently "given".[4] Systems and infrastructures can have political effects, for

---

3   At an early stage, however, it will be unclear what sort of performance and side-effects might be realized. This adds up to a dilemma of knowledge and control (Collingridge 1980), which has become one motivation to do technology assessment at an early stage.

4   An example the "given" character of sociotechnical governance arrangements, often quoted in the STS literature, are the overpasses on Long Island, which continue to "govern" what is possible and what is not possible even after Robert Moses' original intentions became irrelevant (Winner 1980). Their designer, New York city architect Robert Moses, created them to keep New York's black and poor whites (who had to use

example the material unification of a country like the Netherlands (Schot et al. 2003); see also Anderson (1991) for the role of sociotechnical regimes and the idea of a national community. The socio-technical landscapes in/of our societies are like a constitution, even if not drawn up by a constitutional assembly. This includes the disciplining (of actors) necessary to maintain them and have them develop in certain ways. Systems like mobile telephony, including infrastructure as well as evolving customs and rules of use, are further examples of emerging socio-technical regimes which function as a *de facto* constitution.[5]

For emerging sciences and technologies it is not yet clear what their possible sociotechnical constitutional effects might be, but one can anticipate, based on an understanding of dynamics of technological change and its embedding in society (Rip and Te Kulve 2008). This requires what one might call non-linear thinking, especially for technologies like nanotechnology which are enabling technologies. That is, nanotechnology delivers new materials and components to help create better devices and systems, and it is the latter which deliver the desired functionalities, and thus show sociotechnical agency. Thus, nanotechnology is said to just improve performance, and sometimes allow new functionalities (e.g. dirt-repellent surfaces), and should therefore not be an object of concern about its societal effects. Still, nanotechnology can lead to major differences, because certain thresholds may be passed. For example, when RFID (Radio Frequency Identification Devices) becomes cheaper and smaller thanks to nanotechnology, and thus more widely usable as well as better implantable, all products can be traced individually and an "Internet of Things" becomes possible, as well as implants become easy and almost natural, leading to a view of the implantable and thus "readable" human. All this is still to come, but it is being discussed already and may lead to measures and arrangements. One could call this anticipatory governance (Barben et al. 2007). In fact, there is an anticipatory component to all governance (Rip 2006).

The role of technology in governance is one of solidifying arrangements by embodying them in material form. As Pels et al. (2002: 2) phrase it: ".. the performative and integrative capacity of 'things' to help make what we call society." In the case of the overpasses on Long Island discussed in note 4, certain governance modes were delegated to the things, which then did their governing job without b eing

---

busses at the time, the 1920s and 1930s) away from the beaches and parks he had created on Long Island. He tried to create a material constitution for his preferred social order, and while it may have worked for a time, this particular constraint on behaviour has become irrelevant now that every American can use a motor car.

5   This is an Actor-Network Theory notion, cf. how Latour (1991), for similar reasons, speaks of a "Parliament of Things". See also Verbeek (2006) on the morality of artifacts.

recognized for it. Since actual nano 'things' are still (mostly) in the future, such delegation is not possible. But there are expectations, and these can solidify into a forceful societal agenda which will govern strategic choices. One might call this "delegation to the future", and one can definitely see such "delegations" occurring in the domain of nanotechnology.

De facto Risk Governance in the Domain of Nanotechnology

Before tracing the emergence of a risk governance agenda, it is important to note that 'nanotechnology' is an umbrella term, covering quite different scientific and technological developments which are similar only in that they work at the nano scale. In policy making and to some extent in media coverage and public perception, it is the umbrella term that is used, so that differences are black-boxed even where they would be relevant.[6] In the risk debates, the reference is often just to nanotechnology, while the actual concerns, as well as present studies, are about nano-particles (and engineered free nano-particles at that). A reconstruction of the emergence of a forceful agenda will have to take this into account, and maybe explain the focus on nano-particles.

For a reconstruction of the evolving risk governance debate and the resulting *de facto* agenda, I can build on a study by Van Amerom and Rip, based on a comprehensive study of documents (up till 2006) and on interviews and participant observation in relevant meetings, and partly reported in Rip and Van Amerom (2009). They were interested in societal and *de facto* agenda-building as the key phenomenon rather than the traditional focus in agenda-building studies on one single arena and what happens inside that arena. Societal agenda-building is a multi-arena process, without there being a clear authority deciding on the agenda. Kingdon (1984) provided the starting point for their analysis, because of his discussion of policy entrepreneurs and their skills, their networks, and on how they can act on policy windows and other opportunities to forge a new, or change the existing, agenda. This converges with a point made in sociology: "Arenas and fora, and the various issues discussed and addressed there, [which] thus involve ... political activity but not necessarily legislative bodies and courts of law" (Strauss, 1978: 124). Such (always partial) entanglements can become locked-in into a forceful agenda, and then lead to path dependencies (Rip et al. 2007).

---

6   There is a *de facto* governance element involved in such processes: some terms become forceful exactly because they remain blackboxed.

Figure 1, reproduced from Rip and Van Amerom (2009), maps the emerging paths in the evolution of the debate and activities and strategies. Time is on the horizontal axis, and the visualisation of the developments starts with promises about application of nanoparticles as voiced around 2000 and taken up by researchers and firms. The vertical axis comprises ongoing practices of production and use of nanoparticles, then meso-level activities of collective organisations (and of research and regulation), and macro-level societal debate. While there was already a general idea about the promises of nanotechnology, linked to the establishment of the USA National Nanotechnology Initiative in 2000, and some concerns were voiced based on speculative scenarios about run-away nano-robots, macro-level debate only started for real with a Canadian-based NGO (ETC group)'s early warning about risks of nanoparticles and the proposal of a moratorium on their development. This was resisted by nano-enactors, but was listened to, during 2003, by governance actors like the European Parliament and the UK government (see for details Rip and Van Amerom (2009)).

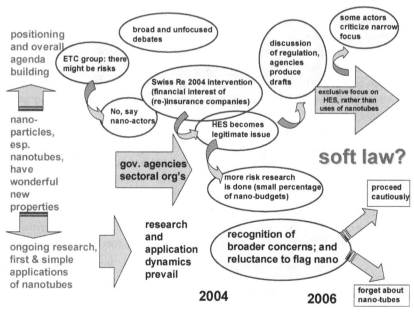

Fig. 1   Evolving paths in the 'landscape' of nanoparticles and their risk

For my analysis of *de facto* governance, I highlight two points visible in the further developments. The first is that arenas overlap, and that actors, in practice, are not limited to their formal roles. Government actors with regulatory responsibility (especially those who are pro-active) attend meetings and generally, take part in a variety of arenas where informal societal agendas are built. Industrial actors mingle with other kinds of actors, especially if a somewhat neutral space is provided. After the intervention in the debate, in 2004, by one of the main re-insurance companies Swiss Re, with a report arguing that carbon nanotubes might create risks similar to those of asbestos, risks of nanoparticles became a legitimate topic. A subsequent meeting organized by Swiss Re and the International Risk Governance Council (IRGC) in Zürich in July 2006 was an occasion for informal interactions.[7] The meeting discussed a (potentially authoritative) report on risk governance of nanotechnology, authored by Renn (a risk and public deliberation scholar) and Roco (of the USA National Nanotechnology Initiative). Governmental and industry actors from across the world attended, as well as NGOs and scientists and scholars studying nanotechnology in society. Dedicated workshops and mingling in the corridors allowed interaction and recognition of positions of other actors (and thus some learning). Very visible was the recurrent anticipation on public reactions (in co-evolutionary terms, this can be described as possibly leading to selection before-the-fact). Clearly, the traditional distinction between formal agenda-building by authoritative (policy) actors, and informal societal *de facto* agenda-building becomes blurred. While one can still recognize, with Shibuya (1997), that for a risk issue to rise on the formal agenda, it needs to be taken up in both formal and informal agenda-building processes, articulation and prioritization processes are clearly not separate.

The second point is how actual soft law, like the recent voluntary reporting schemes of UK Defra and US EPA (cf. Kearnes and Rip 2009), is not just a matter of a new government initiative. It is prepared through actors moving in new directions. As is visualized in Figure 1, such actors can be firms which realize they have to proceed cautiously, and possibly assure credibility by being more transparent. Or regulatory actors recognizing there are openings for regulatory action, but do not know exactly how to proceed. It is the combination of the two which creates a situation where soft law can be envisaged. And even then, there may not be much

---

7   The IRGC is a private not-for-profit foundation, based in Geneva, "to support governments, industry, NGOs and other organizations in their efforts to understand and deal with major and global risks facing society and to foster public confidence in risk governance." (quoted from Renn & Roco 2006, p. 5) A conference report is available from Swiss Re Centre for Global Dialogue.

receptivity. Firms are reluctant to start reporting if they do not know what will be done with such data.[8] As Djelic and Andersson (2006: 378) note for transnational governance, "Soft rules are generally associated with complex procedures of self-presentation, self-reporting and self-monitoring," and may thus lead to more organizing rather than less.

Interestingly, some companies do take initiatives. Codes of conduct are formulated (see further section 3), and a Risk Framework for Nanotechnology has been put forward by the unusual alliance of a big chemical firm (DuPont) and a non-profit group (Environmental Defense). The alliance was announced in June 2005.[9] Their eventual risk framework, published in June 2007 after wide consultation, is a substantial contribution, even if the alliance has come in for criticism by (other) NGOs and trade unions.[10] In the Executive Summary (p. 7), the authors actually note the link with government regulation:

> We believe that the adoption of this Framework can promote responsible development of nanotechnology products, facilitate public acceptance, and support the development of a practical model for reasonable governmental policy on nanotechnology safety.

In other words, actors now contribute to evolving governance arrangements in a reflexive manner.

The implications of this discussion of risk governance agenda building are double. First, that risk assessment in the real world and risk management and regulation are part of larger dynamics, are shaped by it, and their effects (their "success") are partly determined by these broader dynamics. One reason for the dominance of

---

8   By July 2008 only nine companies had registered with the Defra scheme and EPA had received four submissions under the basic program (and commitments from 12 more companies) whilst no company has agreed to participate in the in-depth program. Interestingly, some branch organizations recognizing the importance of the scheme for the credibility of the nanotechnology sector, tried to push their members to participate (see Kearnes and Rip 2009).

9   The two partners had a sense of the historical importance of their attempt when they announced it in an article in *Wall Street Journal*, June 14, 2005, under the title 'Let's Get NanoTech Right'. This echoes an earlier claim about how to handle nanotechnology: "Let's get it right the first time!" Cf. introduction, Roco & Bainbridge 2001 http://www.wtec.org/loyola/nano/NSET.Societal.Implications/

10 Partly in response, an "international coalition of [seven] consumer, public health, environmental, labor and civil society organizations spanning six continents called for strong, comprehensive oversight" of nanotechnology and nanomaterials. Their text has a strong precautionary thrust; "voluntary initiatives are not sufficient". Quoted from the press release, August 1, 2007 (www.nanowerk.com/news/newsid=2306.php).

broader dynamics is the uncertainty about toxicity and exposure of nanotechnology materials (cf. Bowman & Hodge 2006, Dorbeck-Jung 2007). The point is general, however. The fate of risk assessments (i.e., their uptake and their 'translation') is not determined by their own "internal" quality, but by their evolving contexts which are influenced by other/earlier risk assessments and debates, in this case on genetically modified organisms, and earlier, on nuclear energy. Interestingly, in both earlier cases a storyline of escape of modified micro-organisms and a run-away nuclear reactor occurred which returned in the concern about nano-robots getting out of hand. Similarly, regulation is only one element in a range of governance activities and arrangements, which all operate at the interface between nanotechnology and policy/society and add up to a governance "landscape" (Kearnes and Rip 2009). A key element of this landscape, and in a sense a precondition for regulation, is how a *de facto* risk agenda emerged and shaped responses and interactions.

Second, the actions and reactions building up a socio-technical agenda, which solidifies and then shapes further actions and choices, create patchwork governance arrangements rather than a coherent system. This is clear in the way voluntary reporting and other soft law approaches are progressing (haltingly), as well as in the potential uptake of the Dupont & Environmental Defense Risk Framework and the critical reactions to it. Such patchwork arrangements may well allow nanotechnology innovations to continue, and in that sense considered to be productive. They may turn out to be inadequate, however, when something untoward happens, for example when an unusual toxic effect surfaces. The only politically viable response then is to clamp down on nanotechnology in general and introduce harsh precautionary measures.[11] This is not an argument against patchwork governance arrangements, but an indication of inevitable trade-offs.

Discourse and Practice of Responsible Development of Nanotechnology

Whereas the reference to risk, and thus to possible regulation, created some coherence in the evolving patchwork, there is also more open-ended *de facto* governance occurring around nanotechnology, linked to phrases like 'responsible development'

---

11   This is actually one of the three scenarios developed by Douglas Robinson for a Constructive Technology Assessment workshop on responsible innovation in nanotechnology, December 2007. See for the methodology Robinson (2009).

and 'responsible innovation', and in the USA 'responsible stewardship'.[12] The implications are rarely spelled out systematically, but the thrust can be captured in this quote from the US National Research Council:

> Responsible development of nanotechnology can be characterized as the balancing of efforts to maximize the technology's positive contributions and minimize its negative consequences. Thus, responsible development involves an examination both of applications and of potential implications. It implies a commitment to develop and use technology to help meet the most pressing human and societal needs, while making every reasonable effort to anticipate and mitigate adverse implications or unintended consequences. (National Research Council 2006, p. 73)

Clearly, further development of nanotechnology is the main goal, but openings are created for considering "adverse implications or unintended consequences" and perhaps doing something about them. This may invite nano-promoters to consider broader issues at an early stage, and allow other actors to raise questions about present directions of development.

European Commission documents on nanotechnology often refer to responsible innovation, and recently, a further step was taken by preparing and publishing a code of conduct for nanoscience and nanotechnology (N&N) research.[13] The restriction of the code to 'research' was necessary, because of the limited remit of the European Commission in this respect, but the code is broader, and refers also to public understanding and the importance of precaution. There are explicit links to governance: the guidelines "are meant to give guidance on how to achieve good governance", and when this is further specified, there is this interesting item:

> Good governance of N&N research should take into account the need and desire of all stakeholders to be aware of the specific challenges and opportunities raised by N&N. A general culture of responsibility should be created in view of challenges and opportunities that may be raised in the future and that we cannot at present foresee.

A "general culture of responsibility" cannot be created by the European Commission, of course, but they clearly see themselves as contributing to such *de facto* governance.

---

12 See for example the proposal, in California, for a Nanotechnology and Advancement of New Opportunities (NANO) Act by Rep. Honda (D- San José): "The NANO Act requires the development of a nanotechnology research strategy that establishes research priorities for the federal government and industry that will ensure the development and responsible stewardship of nanotechnology." Compare http://www.house.gov/apps/list/press/ca15_honda/NanoAct2008.html

13 C(2008)424 final, 7 February 2008: Commission Recommendation on a code of conduct for responsible nanosciences and nanotechnologies research.

USA and European government actors are not alone in pushing 'responsible development'. There are now also codes of conduct (specifically for nanotechnology) formulated by firms like BASF to address the corporation's responsibilities to "our employees, customers, suppliers and society but also towards future generations",[14] and similar statements, for example by Degussa (now Evonik). [15] Recently, the Swiss retail industry went through the exercise of formulating a code.[16] Then also, there is the recent initiative toward a 'Responsible Nanotechnologies Code', led by a group consisting of the UK Royal Society, an NGO (Insight Investment), the Nanotechnology Industries Association, and supported by a network organised by the UK Department of Trade and Industry.[17] The proposed code goes much further than safe handling of nanotechnology, but it is not clear if and how it will be taken up. Negotiations about a final text that can be made public are still going on.

There has been criticism of codes of conduct as being bland (though not all of them are), and not specifying sanctions. Even so, they create openings for accountability. This does imply that it depends on others willing to call nano-enactors into account whether there will be real effects of a code. While the discourse of responsible development will have implications for practices, broader issues referred to in the discourse will often be backgrounded. Actual operationalizations of 'responsible innovation' tend to focus on risk issues, transparency and some public dialogue. And in the case of industry, also as a responsibility for safe handling of nano-production and nano-products.

While the recent emergence of Codes of conduct (actual and proposed) already indicates distributed governance, they should be seen as the tip of an iceberg of anticipatory and reflexive actions and interactions which fill up the gap between further development of nanotechnologies and actual and possible responses by society. There is a plethora of activities and gatherings in the nano-world with governance elements and/or implications, with explicit or implicit reference to responsible development. One can see them as emerging practices of discussion, deliberation, negotiation and participation.

---

14   *Code of Conduct Nanotechnology.* http://corporate.basf.com/en/sustainability/dialog/
     politik/nanotechnologie/verhaltenskodex.htm?id=VuGbDBwx*bcp.ce. (accessed 4
     March 2008).

15   Degussa's website on nanotechnology has an item to this extent on responsibility (www.
     degussa-nano.com/nano).

16   Interessengemeinschaft Detailhandel Schweiz. 2007: *Code of Conduct: Nanotechnologies.*
     http://www.cicds.ch/m/mandanten/190/download/CoC_Nanotechnologien_
     final_05_02_08_e.pdf.

17   See the May 2008 update at www.responsiblenanocode.org.

In some cases responsible development is a secondary effect. Definitions and standards for nanotechnology are of immediate importance for coordination among firms, but they will also be used to indicate the scope of regulation and of soft law like voluntary reporting. The International Standards Organization (ISO) has established working parties, and its standards are sometimes called "soft governance".[18] They are voluntary, but recognized as important and authoritative because of the process leading to them (expert working parties and wide consultation). Actors/ stakeholders refer all the time to ISO standards and the working parties because they expect resolution of uncertainty through them. OECD has got involved as well and also looks at risks and public engagement; its working parties have a certain status and are expected to come up with authoritative conclusions. UNESCO has invested in a report on ethics of nanotechnology (UNESCO 2006). Also, dedicated groups or associations are established, for example the International Council On Nanotechnology (ICON) collecting many stakeholders (but almost no NGOs). A web of activities and interactions results, and actors in the nano-world can refer to it to show that responsible development is taken seriously.

Policy actors and nanotechnology spokespersons from industry try to keep abreast of what is happening and thus monitor the evolution of the discourse and the positioning of the various actors and groups. The director of the nanotechnology R&D programme in the European Commission's 6th and 7th Framework Programmes (until September 2008), Renzo Tomellini, often showed a slide, giving an overview (Figure 2). The fact that he shows it, and updates it, is just as important as the content of the slide.[19] The link with responsible development is clear in his mind. When (in meetings in 2007) presenting data on public opinion about nanotechnology, he was happy to note that the European public is more positive than the North-American public. "So we have done a good job. But this trust in us also creates a responsibility to make sure that nanotechnology is developed in the right way."

---

18 See presentation by Peter Hatto (ISO) at the International Dialogue meeting, Brussels, March 2008. (Tomellini and Giordani 2008)

19 There is now an official version (without the clouds in the middle), published on Cordis. ftp://ftp.cordis.europa.eu/pub/nanotechnology/docs/a-interactions-global.pdf

Main International Fora and Initiatives on Nanotechnology

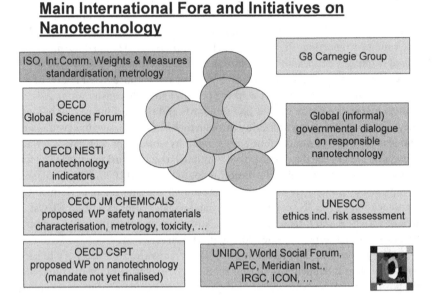

ISO, Int.Comm. Weights & Measures standardisation, metrology

G8 Carnegie Group

OECD Global Science Forum

OECD NESTI nanotechnology indicators

Global (informal) governmental dialogue on responsible nanotechnology

OECD JM CHEMICALS proposed WP safety nanomaterials characterisation, metrology, toxicity, ...

UNESCO ethics incl. risk assessment

OECD CSPT proposed WP on nanotechnology (mandate not yet finalised)

UNIDO, World Social Forum, APEC, Meridian Inst., IRGC, ICON, ...

Fig. 2   Tomellini's overview of activities in the nano-world (version summer 2007)

A further activity that is particularly interesting for what it might do (rather than what it does at the moment) is the International Dialogue on Responsible Development of Nanoscience and Nanotechnology, set in motion by Mike Roco (US National Nanotechnology Initiative) and Renzo Tomellini (European Commission's Nanotechnology Program) in 2004, and perhaps now getting a momentum of its own. The idea was and is to have informal interactions between government officials and other actors in the nano-world, with reference to responsible development as one reason why coordination is important. After the first meeting in Alexandria, Virginia (Meridian Institute 2004), there was a delay because of political difficulties, but then meetings were held in Tokyo (2006) and Brussels (2008), with a next meeting planned in Russia. Such meetings offer reporting on developments, including ethical and social aspects (ELSA) and experiences with public dialogues, but are also spaces for interaction. Their advantage is that they can be inclusive: there is no official mandate or link to an authority, so no actual or symbolic barriers to participation.

The key point to draw out of this mapping of activities is how a variety of actors start discussing responsible development of nanotechnology, refer to it, and develop relevant activities, and how the discourse shows some convergence. This may be the beginning of a shift in governance, driven by prudence and some good intentions, as well as need to maintain legitimacy. Policy actors may be involved, but often interactively rather than stipulating a governance arrangement, and they build on receptivity to the discourse on responsible development that is clearly out there. Thus, there is some *de facto* governance of nanotechnologies. A subsequent question then is whether this is specific to the nature and situation of nanotechnologies, or whether it reflects a general shift in governance, in the direction of reflexive modernization (Beck 1992; Beck et al. 2003)? The latter will be the case, definitely, but nanotechnologies offer a "lead" domain where the shift is visible, and is, in a sense, experimented with.

An Overarching Pattern?

There are more activities and emerging structures to be mapped, not necessarily specific to nanotechnologies, but taken up in earnest there. Upstream public engagement, including citizen conferences after the Danish model, is one example (Kearnes et al. 2006), the interest in having ELSA (Ethical, Legal and Social Aspects) included in the big nano-research funding programs in the USA and Europe is another example.[20] Together with codes of conduct of various status, and emerging soft law with a precautionary flavor, these fill up the gap between promotion and control of nanotechnologies. Thus, the traditional governance arrangement of new technologies is shifting. The new activities and interactions are expected, and this will orient (enable and constrain) strategies, actions and interactions, and will be seen as legitimate.

We see an emerging overarching pattern, but its strength and eventual shape are still unclear. A key component of the pattern is anticipatory governance, and in particular, the consideration, at an early stage, of eventual societal embedding, up to accepting some responsibility. Will what is still mainly discourse become a practice, and a practice of *de facto* governance at that? To explore this question further, I will mobilize insights from STS and economics of innovation, because there is structural similarity with patterns that have emerged in technological development *per se*.

---

20 The USA National Nanotechnology Initiative has funded two big Centers for Nanotechnology in Society, and some smaller units. NanoNed (www.nanoned.nl).

Evolutionary economics and sociology of technological development have identified (and theorized) so-called 'trajectories' of development at two levels. There
are paths of successive specific designs and products, related to what Dosi (1982)
called the 'technological paradigm' which shapes expectations about productive
directions of development, and thus defines requirements. Nelson and Winter (1977:
56) refer to technicians' beliefs about what is feasible or at least worth attempting.

> When such beliefs and corresponding design and development practices are entrenched,
> one can speak of a 'technological regime' determining technological decisions. An
> example is how the advent of the DC3 aircraft in the 1930's defined a particular tech-
> nological regime: metal skin, low wing, piston powered planes. Engineers had some
> strong notions regarding the potential of this regime. For more than two decades
> innovation in aircraft design essentially involved better exploitation of this potential;
> improving the engines, enlarging the planes, making them more efficient.

Then, there are broad design heuristics or guidelines like mechanisation (since the
19th century), and automation (definitely from the 1950s on), what Nelson & Winter
(1977) called a 'natural trajectory' and I will call a second-order trajectory, to bring
out the relation to technology-specific and thus first-order trajectories.

In specific developments in biotechnology, genomics, stem cells, and nanotechnologies, one can find first-order trajectories, some still emerging others already
established. For micro- and nanotechnologies, there is a second-order trajectory
of miniaturization. Such overall requirements on development of new technologies
constitute a governance pattern. My additional point now is that the requirements
need not be limited to those that specify technical performance.[21]

At present, for all newly emerging technologies, one sees attempts to include
societal aspects and to anticipating on embedding in society. Already because of
credibility pressures, such anticipation functions as a soft requirement on specific
developments. Thus, one can hypothesize that a further second-order trajectory
may emerge, which could be labeled 'working towards adequate societal embedding' of technology. There is no guarantee that it will indeed become a trajectory
of technological development even if policy actors are keen on it, as is visible in
the discourse of 'responsible innovation'. I note that, just as with mechanization
and automation, the second-order trajectory of 'working towards adequate societal
embedding' need not be taken up in all concrete developments for it to function as
a governance arrangement. But there must be sufficient actual uptake, and broad
reference to anticipation on societal embedding to make it a second-order trajec-

---

21 While used in another context, the notion of meta-rules of the game (Djelic and An-
    dersson 2006: 385, 391) indicates a similar phenomenon.

tory. This emerging second-order trajectory is not specific for nanotechnology, but it is there where the indicators are strongest – it is the lead technological domain for the trajectory.

Part of such a second-order trajectory seems to be in place already: the inclusion of EHS (Environmental, Health, and Safety) aspects in technological developments at an early stage. This actually builds on what one could call an earlier internalization of requirements from the selection environment: the chemical industry's Responsible Care programme in the 1990s (King and Lenox 2000). It is significant that firms presenting a code of conduct (or similar) for nanotechnology are chemical companies. The focus on EHS may create a lock-in: In nanotechnology, as well as for other new technologies like GMO, adequate societal embedding will quickly be attention to reduced to EHS aspects. Some actors, however, like Degussa (a chemical company) do emphasize the importance of responsibility and dialogue, and attempt to interact with new actors like critical NGOs. It is uncertain whether such operationalizations of broader anticipation

An indirect indicator of internalization is that funding agencies start creating special programs on ELSA and societally responsible innovation.[22] In a sense, funding agencies are "third parties": they do not develop (nano)technology themselves, but through their actions will influence developments.[23] Capital providers like banks and pension funds (and perhaps even venture capitalists looking further than immediate return on investment) might play such a third-party role as well when they would introduce requirements of responsible development. When funding agencies and other sponsors of research and development actually require anticipation on adequate societal embedding, nano-enactors have to develop relevant competencies, and this will contribute to solidification of the arrangement.

---

22 In the Netherlands, such a program has just started; it focuses on (a) advanced (emerging) technologies and (b) sociotechnical system transitions (www.nwo.nl/mvi). In Norway, the theme is ELSA of biotech, nanotech and neurotech (www.forskningsradet.no). In both cases interaction between social science and humanities and on the other hand, science and engineering is an important requirement. In the UK, the Engineering and Physical Sciences Research Council (EPSRC), established a Societal Issues Panel in 2006, and experimented with dialogue. In the words of a participant observer: "An emergent sociotechnical imaginary that takes 'societal issues' not as an obstacle but as an active contributor to framing the work of the research councils." (Doubleday 2008)

23 Analytically, the importance of such "third parties" taking initiatives is that they can break through waiting games and other impasses that occur often in two-party games (Scharpf 1997). An example of such a breakthrough is the intervention of re-insurance company Swiss Re in the risk debate (see section 2).

In Conclusion

Broadening the notion of governance has enabled me to discuss *de facto* govern-ance, in general and how it occurs in the domain of nanotechnology. There were attempts at intentional governance, addressing uncertainties around an emerging technology, but their fate had to be understood against the backdrop of evolving *de facto* governance. The importance of societal agenda building, through interactions of actors and their strategies, was clear. For risk governance, our understanding of such emerging patterns might be used to create scenarios of possible futures, to improve anticipatory governance, or at least increase reflexivity. The discourse of responsible development also showed a mix of attempts at intentional governing (as in the case of the European Commission's Code of Conduct for N&N research), actors' initiatives and emerging patterns in a web of interactions creating orderings in the world of nanotechnologies. These are general features of *de facto* governance.

Striking in the emerging *de facto* governance of nanotechnologies was the role of anticipation. Actors anticipate on possible futures as well as reactions from other actors. This is more than prudence: newly emerging technologies like nano-technologies create openings (opportunities as well as some concerns) which are uncertain by definition. The future becomes a reference point, even if it is unknown. One can speak of the "shadow of the future", in the same vein as Scharpf (1997) talks of the "shadow of hierarchy". Scharpf uses the metaphor of 'shadows' more widely than just for hierarchy, e.g. "in the shadow" of the majority vote (at p. 191) or "in the shadow" of a statute (at p. 202), without being explicit about the actual mechanisms and dynamics (other than his reference to anticipated reactions when discussing decision making in bureaucracies, at p. 200). For Scharpf, a key notion is 'authority structure' and how this can be referred to and thus have effects in an indirect way. Thus, for new and emerging technologies like nanotechnologies, "the future", when articulated in more or less forceful societal agendas and expectations about responsible development, functions as an authority structure, and thus casts its shadow on choices and actions here and now.

De facto governance is distributed, almost by definition, and cannot be easily shaped from a central point. This introduces ambivalence in the role of governance actors like governments. They have to give up on the assurance of governability because lots of things are outside their power and influence. At the same time, there is *de facto* governability: social orders are there, and are (somewhat) effective – even if the direction may not be ideal. Actually, there is co-evolution of intentional and *de facto* governance anyhow, and governance actors might then see their role as one of modulating this co-evolution (Rip 2006).[24] When UK Defra and US EPA went for voluntary reporting, they hoped they could draw on a sense of responsibility with the firms, but clearly they were too early to effectively modulate: the evolution at the other side had not progressed sufficiently (the situation might be different in continental Europe). The International Dialogue discussed in section 3 started from the other side: while government actors are involved, they do not govern, but create a space in which *de facto* governance might be stimulated.

Modulation of co-evolution occurs as well in the public dialogues, in the creation of voluntary codes and responses by civil-society actors, and in technology assessment interactions visible in the world of nanotechnologies. It is explicitly mentioned in the recent calls for 'midstream modulation' (Fisher et al. 2006) and midstream public engagement (Joly and Rip 2007). The work of governing is distributed, and may become partially internalized when regimes stabilize. The possibility of a second-order trajectory where working towards adequate societal embedding is a requirement on ongoing technological developments is a particular, and particularly interesting, case.

Even without a fully fledged second-order trajectory, nano-enactors already take initiatives, by themselves or stimulated by third parties. This works out differently in the various domains under the umbrella term 'nanotechnology'. For new materials, chemical companies have relevant competencies because of the earlier (and continuing) Responsible Care Programme, and they feel credibility pressures. It is in this sector that firms have come up with nanotechnology codes of conduct. Micro-electronics firms, on the other hand, who do a lot of work at the nano-scale, are far removed from end users (even if they try to create some visibility, as with labels Intel inside" on laptops). There are discussions, for example about RFID and about ambient intelligence, systems enabled by nanotechnology, but there other firms take the lead.[25] Big incumbents have the resources to be pro-active, but do not always rise to the occasion. In bionanotechnology, the third main domain of nanotechnology, the big pharmaceutical companies are interested, but tend to wait for small firms coming up with nano-enabled innovations like diagnostic devices and drug delivery. For small firms, their first concern is to survive, broader anticipation is a luxury. Third parties like insurance companies may be able to (sometimes

---

24 This is similar to what Andrew Dunsire (1996) has called "collibration" – an intervention to shift the preexisting balance between countervailing forces. Institutional arrangements that have the effect of strengthening one or weakening the other of these forces will require much less energy than institutions that would have to stop an unopposed force. (Quoted from Scharpf 1997: 182).

25 Firms like Philips and Siemens who used to cover both sectors have now divested their micro-electronics development and production.

inadvertently) modulate productively. The other input in de facto governance of nanotechnologies would come from interactions across the product-value chains. The first signs of this are linked to health and environmental issues.

Thus, *de facto* governance is not blind. It is shot through by attempts at shaping, and by their residues, up to somewhat stabilized regimes around nanotechnologies, and newly emerging technologies more generally. The question of the quality of such governance, e.g. governability, legitimacy and directions that are pushed, is important but remains difficult to handle since no one actor is specifically responsible. However, as soon as regimes and second-order trajectories appear, these offer entrance points for critical evaluation and perhaps attempts at changing them – by modulation.

Bibliography

Anderson, B. (1991). *Imagined communities. Reflections on the origin and spread of nationalism* (revised ed.). London: Verso.

Barben, D., Fisher, E., Selin, C., & Guston, D. H. (2007). Anticipatory governance of nanotechnology: Foresight, engagement, and integration. In E. J. Hackett et al. (Eds.), *The handbook of science and technology studies* (3rd ed.). (pp. 979-1000). Cambridge, MA: MIT Press.

Basic Books National Research Council. (2006). *A matter of size: Triennial review of the national nanotechnology initiative.* Washington, D.C.: National Academies Press.

Beck, U. (1992). *Risk society. Towards a new modernity.* London: Sage Publications.

Beck, U., Bonss, W., & Lau, C. (2003). The theory of reflexive modernization. Problematic, hypotheses and research programme. *Theory, Culture & Society, 20*, 1-33.

Bowman, D. M., and Hodge, G. A. 2006: Nanotechnology: mapping the wild regulatory frontier. *Futures* 38(9), 1060-73.

Braithwaite, J., & Drahos, P. (2000). *Global business regulation.* Cambridge: Cambridge University Press.

Collingridge, D. (1980). *The social control of technology.* London: Frances Pinter.

Djelic, M.-L., & Andersson, K. S. (Eds.). (2006). *Transnational governance. Institutional dynamics of regulation.* Cambridge: Cambridge University Press.

Dorbeck-Jung, B. (2007). What can prudent public regulators learn from the United Kingdom Government's nanotechnological regulatory activities? *Nanoethics, 1,* 257-270.

Dosi, G. (1982). Technological paradigms and technological trajectories: A suggested interpretation of the determinants and directions of technical change. *Research Policy, 11,* 147-162.

Doubleday, R. (2008). *No room for doubt: Public engagement, science policy and democracy at the UK's Engineering and Physical Sciences Research Council.* Paper presented at the meeting of the Society for Social Studies of Science and the European Association for the Study of Science and Technology. Rotterdam.

Dunsire, A. (1996). Tipping the balance: Autopoiesis and Governance. *Administration and Society, 28*, 299-334.

Fisher, E., Mahajan, R. L., & Mitcham, C. (2006). Midstream modulation of technology: Governance from within. *Bulletin of Science, Technology and Society, 26*(6) 485-496.

Hanf, K., & Toonen, T. A. J. (1985). *Policy implementation in federal and unitary systems. Questions of analysis and design.* Dordrecht, Boston, Lancaster: Martinus Nijhoff Publishers. NATO Advanced Science Institutes Series, D23.

Joly, P.-B., & Rip, A. (2007). A timely harvest. *Nature, 450,* 308.

Kearnes, M. B., Macnaghten, M., & Wilsdon, J. (2006). *Governing at the nanoscale: People, policies and emerging technologies.* London: Demos.

Kearnes, M., & Rip, A. (2009). The emerging governance landscape of nanotechnology. In S. Gammel, A. Losch & A. Nordmann (Eds), *Jenseits von Regulierung: Zum politischen Umgang mit Nanotechnologie.* Berlin: Akademische Verlagsanstalt, forthcoming.

King, A., & Lenox, M. J. (2000). Industry self-regulation without sanctions: The chemical industry's responsible care program. *The Academy of Management Journal, 43*(4), 698-716.

Kingdon, J. W. (1984) *Agendas, alternatives and public policies.* Boston/Toronto: Little, Brown and Company.

Kooiman, J. (2003) *Governing as governance.* London: Sage Publications.

Latour, B. (1991). *Nous n'avons jamais été modernes. Essai d'anthropologie symétrique.* Paris: La Decouverte.

Meridian Institute & National Science Foundation. (2004). *Report: International dialogue on responsible development of nanotechnology.* Washington, D.C.: Meridian Institute.

Mintzberg, H. (1994). *The rise and fall of strategic planning.* New York, NY: Simon and Schuster.

Nelson, R., & Winter, S. (1977). In search of a useful theory of innovation. *Research Policy, 6,* 36-76.

Ostrom, E. (1990). *Governing the commons: The evolution of institutions for collective action.* Cambridge: Cambridge University Press.

Pels, D., Hetherington, K., & Vandenberghe, F. (2002). The status of the object. Performances, mediations, and techniques. *Theory, Culture & Society, 19*(5/6), 1-21.

Pressman, J.L., & Wildavsky, A. (1984). *Implementation. How great expectations in Washington are dashed in Oakland* (3rd ed.). Berkeley, CA: University of California Press.

Renn, O., & Roco, M. (2006). *Nantechnology risk governance.* Geneva: International Risk Governance Council, June 2006. White Paper #2.

Rip, A., Misa, T. J. & Schot, W. (Eds.). (1995). *Managing technology in society. The approach of constructive technology assessment.* London: Pinter Publishers.

Rip, A., & Groen, A. (2001). Many visible hands. In R. Coombs, K. Green, V. Walsh & A. Richards (Eds.), *Technology and the market. Demands, users and innovation* (pp. 12-37). Cheltenham: Edward Elgar.

Rip, A. (2006). A co-evolutionary approach to reflexive governance – and its ironies. In J.-P. Vos, D. Bauknecht & R. Kemp (Eds.), *Reflexive governance for sustainable development* (pp. 82-100). Cheltenham: Edward Elgar.

Rip, A., Robinson, D. K. R., & te Kulve, H. (2007). *Multi-level emergence and stabilization of paths of nanotechnology in different industries/sectors, paper prepared for International Workshop on Paths.* Berlin, 17-18 September 2007.

Rip, A., & te Kulve, H. (2008). Constructive technology assessment and sociotechnical scenarios. In E. Fisher, C. Selin & J. M. Wetmore (Eds.), *The yearbook of nanotechnology in society* (1st ed.): *Presenting futures* (pp. 49-70). Berlin etc: Springer.

Rip, A., & van Amerom, M. (2009). Emerging de facto agendas around nanotechnology: Two cases full of contingencies, lock-outs and lock-ins. In S. Maasen, M. Kaiser, M. Kurath & C. Rehmann-Sutter (Eds.), *Deliberating future technologies: Identity, ethics, and governance of nanotechnology*. Heidelberg et al.: Springer, forthcoming.

Robinson, D. K. R. (2009). Complexity scenarios for emerging techno-science. *Technological Forecasting and Social Change, forthcoming*.

Scharpf, F. W. (1997). *Games real people play. Actor-centred institutionalism in policy research*. Boulder, Col.: Westview Press.

Schot, J.W., Lintsen, H.W., Rip, A., & de la Bruheze, A. A. A. (2003). (en mede-redactieleden), *Techniek in Nederland in de Twintigste Eeuw. VII. Techniek en Modernisering. Balans van de Twintigste Eeuw*. Zuthphen: Walburg Pers.

Shibuya, E. (1996). Roaring mice against the tide: The South Pacific Islands and agenda-building on global warming. *Pacific Affairs, 69*, Winter 1997/1997.

Strauss, A. (1978). A Social World Perspective. *Studies in symbolic interaction, 1*, 119-128.

Swiss R. (2004) *Nanotechnology: Small matter, many unknowns*. May. Zurich: Swiss Re. 56.

Tomellini, R., & Giordani, J. (2008). *Report: Third international dialogue on responsible research and development of nanotechnology*. Brussels: European Commission. Retrieved from https://ec.europa.eu/research/industrial_technologies/pdf/policy/report-third-international-dialogue-2008_en.pdf

UNESCO, Division of Ethics of Science and Technology (2006). *The ethics and politics of nanotechnology*. Paris: UNESCO.

Van Kersbergen, K., & van Waarden, F. (2004). Governance as a bridge between disciplines: Cross-disciplinary inspiration regarding shifts in governance and problems of governability, accountability and legitimacy. *European Journal of Political Research, 43*, 143-171.

Verbeek, P.-P. (2006). Materializing morality—design ethics and technological mediation. *Science, Technology & Human Values, 31*(3), 361-380.

Voβ, J.-P. (2007). *Designs on governance. Development of policy instruments and dynamics in governance*. Enschede: University of Twente.

Chapter 5
Constructive Technology Assessment*

As it turned out, I hadn't written a paper on Constructive Technology Assessment that offered a comprehensive discussion. So I had to go for a compilation. It was also an opportunity to add an introduction in which the course of my work on Constructive Technology Assessment could be sketched.

Introduction

In the 1984 Policy Memorandum on Integration of Science and Technology in Society (to which I had contributed when seconded to the Ministry of Education and Sciences at the time), there was a plea for broadening design and development of technology by including more aspects and more actors, and to see this as a constructive way of doing technology assessment (Ministerie van Onderwijs en Wetenschappen 1984). On the basis of the Policy Memorandum, a Netherlands Organization for Technology Assessment (now Rathenau Institute) was established in 1986. One of its first projects was to develop this general plea into an approach, to be called Constructive Technology Assessment (CTA). The Report (Daey Ouwens et al. 1987) had case studies offering diagnoses, theory derived from technology dynamics, and some proposals.[1] The signature feature of CTA, soft intervention based on applied technology dynamics, was in place, even if only programmatically.

---

* Source: A compilation from existing publications, with an introduction about earlier and ongoing work

1 Van den Belt and Rip's contribution to the Report was the basis for presentations and some publications on concepts and possibilities of Constructive TA, and was revised to serve as the background paper for the 1992 international workshop. In parallel, I developed background theory further, see especially my papers in the new and short-

© Springer Fachmedien Wiesbaden GmbH, part of Springer Nature 2018
A. Rip, *Futures of Science and Technology in Society*, Technikzukünfte, Wissenschaft und Gesellschaft / Futures of Technology, Science and Society, https://doi.org/10.1007/978-3-658-21754-9_6

In the subsequent years, I focused on developing theory and conceptualization of technology dynamics and their applications in technology and innovation policy and management, as well as in (Constructive) Technology Assessment. NOTA/ Rathenau had some exploratory projects, and Johan Schot (who wrote up his studies in his 1991 PhD thesis as contributions to an emerging paradigm of Constructive TA) and I, with Tom Misa, were able to convene an international workshop in 1992, devoted to diagnosis of, and approaches to, constructive intervention in technology development and its embedding in society. A selection of the papers was published in 1995 (Rip, Misa and Schot 1995).

During the 1990s, we were able to "do" Constructive TA in a number of domains, in particular social experiments with electric vehicles and "micro-optics", a confluence of technologies to replace, or at least complement, micro-electronics. We also made a first attempt to articulate and position the program of CTA (Schot and Rip 1997). In parallel, I had developed a multi-level model of technological development and its embedding in society, with niches, regimes and sociotechnical landscapes as modes of ordering ongoing practices of technological development (i.e. enabling and constraining them), with different scope and time scale.[2]

While early on, existing insights in technology dynamics were used to identify possibilities for changing, or at least modulating, technological development, this question soon led to new work in technology dynamics, as visible in a series of PhD theses ranging from a study of promising technology (Van Lente 1993) to changing technology (Van de Poel 1998),[3] and also sociotechnical transitions (Geels 2001). We synthesized technology and innovation dynamics literature to identify *loci* for intervention (Rip and Schot 2002). And I continued analysis of co-evolution of technology and society as becoming more reflexive, also as a way to position concrete tools and approaches of technology foresight and technology assessment. (Rip 2002)

---

lived Italian journal RISESST – which did manage to publish articles by Giovanni Dosi, Bruno Latour, Arie Rip (Rip 1992a, b).

2   Cf. Rip and Kemp (1998). The multi-level visualization with arrows indicating technology development over time, now widely used and reproduced in the versions developed by Geels since his (2002) PhD Thesis, was taken out of the submitted (commissioned) manuscript when the editors of the book revised the manuscript, and was published only in Rip (2000). (Cf. also Kemp et al. 2001) The key point, that niches, regimes and landscapes are modes of ordering is not always recognized.

3   There was intentional ambiguity in the titles, and this ambiguity was continued in Deuten (2003) and in a recent PhD thesis on Circulating Images of Nanotechnology (Ruivenkamp 2011).

Thus, the overall philosophy of CTA, soft intervention to broaden technology development "for the better", inspired scholarly studies of evolving technological developments and their embedding in society. When I was asked, in 2003, to set up and conduct a research program on CTA of nanotechnology, a further possibility emerged: using CTA interventions to improve our understanding of technology dynamics as these occur. The goal of the research program was to develop a CTA methodology for a technology which lives on promises (and in that sense is science fiction, so that eventual impacts will be social science fiction), but the publications and PhD theses that came out of the program also contributed important insights in technology dynamics and embedding in society.[4] One example is how the phenomenon of waiting games around novel and uncertain technologies could be linked to dual dynamics of promises (Parandian et al., forthcoming).

CTA works out differently at different stages of technology development, and one can contrast electric vehicles and nanotechnology.[5] The differences also have to do with which level (in the technical hierarchy) is focused on: materials, components and devices (as in advanced technology) or (socio-)technical systems. The main target group of CTA exercises/interventions might shift as well. Up to now, producers (enactors) of technology were key targets, and for good reasons if one wants to influence developments at an early stage. But when looking at socio-technical systems, it is easier to target government agencies, consumers and civil society groups as well. Such actors (including so-called "third parties") have always indirect influence on technology development, but when embedding of technology in society is at stake, they are directly involved.

Soft intervention by CTA actors will have an overall rationale: better technology in (a better) society (as Schot and Rip (1997) phrased it), and/or more reflexive co-evolution of technology and society (cf. Rip 2002).[6] The objectives are more

---

4 Van Merkerk (2007), Robinson (2010), Ruivenkamp (2011), Te Kulve (2011), Selley-Egan (2011). Two more PhD theses, by Parandian and by Krabbenborg, are in the pipeline.

5 The notion of an "early stage" is relative: one might see electric vehicles as being at a late stage of technological development: starting in the late 19th century, surviving in niche applications, now getting new leases on life. But their actual embedding and broader uptake requires further socio-technical innovations, and is therefore in an early stage. This is how social experiments with electric vehicles have been studied (Hoogma 2000, Hoogma et al. 2002).

6 I often refer to a "philosophy" of technology assessment: reduce the costs of learning by trial-and-error which characterizes much of our handling of technology in society, by anticipating future developments and their impacts, and by accommodating such insights in decision making and its implementation. Such anticipation is not limited to commissioned TA studies, but will be part of a societal learning process, in which

modest, though: to work through (1) opening up spaces, including dedicated (and thus orchestrated) spaces for actors to interact and because of these interactions, do better; and (2) by offering analysis and understanding of dynamics and force fields, and tools to support actors in articulating perspectives and strategies.

The tools are "reflexivity tools", even if they must have recognizable instrumental value to the target groups so that these will appreciate the tools and use them. The tools should be of such a kind (and/or so packaged), that when they are used, the user cannot remain purely instrumental, and just work to achieve existing own goals. The user is forced to reflect on the broader situation, and perhaps adjust his goals. Socio-technical scenarios and their use in workshops with different stakeholders are a good example because they introduce complexity and show that this is necessary (Robinson 2009).

The Why and How of Constructive TA
(Excerpt from Rip (2008), pp. 145-150)

At a macrolevel, the co-evolution of science, technology and society can lead, and has in fact led, to stable patterns or regimes (sometimes including explicit governance arrangements). The distinction between technology variation and societal selection (which carries part of the argument about an evolutionary approach) is actually predicated on the existence of an historically evolved regime where technology development is separate from uptake and use (Rip and Kemp, 1998). Variation and selection are institutionally separated – and are bridged by anticipations, experiments, and interactions.

Historically, the separation was part of the industrial revolution. In a further step, during the 19th century, engineers and other technology actors were given a mandate to develop new technologies and confront society with them, as long as this could be presented as progress. This was complemented by the emergence of the idea of 'technology' in general, symbolically linked to ideals of progress (Smith and Marx, 1994). The mandate and its symbolic justification offered a macroprotected space. It was maintained until the 1960s (when Harold Wilson, the then UK Prime Minister, could still proclaim a 'white-hot technological revolution'), but then it started to break down.

---

many actors can participate. One can see this as one operationalization of co-evolution becoming more reflexive.

The existence of the mandate, and the macroprotected space for technological developments, however, had allowed the establishment of institutions and divisions of labour which could not simply be turned back. While shifts do occur, and there are contestations, the overall pattern of a gap between promotion and control and attendant division of labour between promoters of new technology and critical selectors remains strong.

Thus, in our modern societies there is an asymmetry between 'impactors' (those at the source of technological impacts) and 'impactees'. This can be a difference in power, but is always a difference in timing. Initiators of technological development know more, and have invested more, at an early stage, and impactees and spokespersons for society have to wait and in a sense, follow their lead. Collingridge (1980) has identified the knowledge and (public) control dilemma which follows this situation. The problem is already visible within 'innovation journeys' (Van de Ven et al., 1999) as a flexibility dilemma (Verganti, 1999): one has to foreclose options at a moment when not enough is known. In fact, assessment is ongoing, and the challenge is to improve the assessment process by, for example, including more aspects and perspectives at an early stage as Constructive TA proposes (Rip et al, 1995; Schot and Rip, 1997).

The asymmetry has another component: technology developers are insiders and do not necessarily know very much about the 'outside'. However, adoption and diffusion, is up to 'outsiders', who have other interests and expectations. The story of nuclear energy has been one of struggle between insiders and outsiders since the 1960s. On a much smaller scale, the same storyline is visible in the development of cochlear implants for deaf people (Garud and Ahlstrom, 1997), especially when it turned out – unexpectedly for the insiders – that the deaf community was negative about taking deaf people out of their own culture by providing them with implants (Blume, 2000; Reuzel, 2004). Garud and Ahlstrom (1997) suggest that technology developers are working in an 'enactment frame', and look at the world as a challenge or barrier to be overcome (while from the outside actors can be selective according to their various backgrounds). The 'enactment frame' then leads to a concentric approach to product development: get the product right, then look at market and regulation, and afterwards worry about public acceptability. Deuten et al. (1997) have mapped this phenomenon for biotechnology products. As they have shown in a narrative analysis of one of the cases the 'enactment frame' itself evolves, because, for example, other 'niches' emerge (Deuten and Rip, 2000).

For our topic, it is important to note that the asymmetry gives rise to a division of TA labour where insiders articulate 'promotion', and outsiders 'control'. A similar division of labour, at one remove, is visible in the way governments and their agencies handle technology in society. Government technology policy most often focuses on promotion of (selected) technologies, as in the case of stimulating the electronic superhighway. The main challenge for such technology policy is to 'pick the winners', at the level of the society. At the same time, other agencies of the same government are occupied with reducing the human and social costs of the introduction of new technology, for example through safety and environmental regulation. This dichotomy between promotion and control of new technology is thus part of the *de facto* constitution of modern industrialised societies, and is reflected not only in the division of labour between government agencies, but also in cultural and political views, as in the assumption that there will be proponents and opponents to a new technology.

The gap is precariously bridged by a variety of instruments and actions, by existing and emerging nexuses and by division of labour in sociotechnical regimes. It is in this world full of asymmetries and gaps that Constructive TA (Rip et al., 1995; Schot and Rip, 1997) is located. Theoretically – it can be seen as a series of bridging events, and when institutionalised, as a nexus; practically – it organises interactive workshops supported by sociotechnical scenarios; and normatively – there is an overall aim to improve technology and society, and a conviction that increasing reflexivity of ongoing co-evolution will help. Democratisation of technology can be associated with Constructive TA, but that is a possible effect, not a goal. In other words, participation is a means, *not* an end. The key point is to enhance reflexivity through anticipation, feedback and learning (Schot and Rip, 1997), and translate this into actions and interactions.

(..)

There is, by now, a range of approaches to technology assessment. Foresight and future visioning emphasise the open future, and there are now proposals for 'vision assessment' (Grin and Grunwald, 2000). At the other end of the spectrum there is the comparison of existing technological options by firms and R&D institutions in order to select the promising ones. Within the range of approaches, a cluster of approaches and methodologies have been developed and piloted over the last 10-15 years, which emphasise real time interaction and learning. There are various labels, including Interactive TA (Grin and Van de Graaf, 1996; Grin, et al., 1997), Real-Time TA (Guston and Sarewitz, 2002), and Constructive TA (Rip et al., 1995). We prefer the label Constructive TA to emphasise the goal of contributing to the actual construction of new technologies and the way these become more or less embedded in society – rather than simply waiting for the changes and then trying to map possible impacts.

For new and emerging science and technology like nanoscience and nanotechnologies, Constructive TA has to address what can be called its doubly 'fictional' character: many of the envisaged uses of nanotechnology are still science fiction, and so the study of possible impacts is social science fiction. In our specific situation – the TA and Societal Aspects of Nanotechnology Program, within the Dutch R&D consortium NanoNed – there is an opportunity: being part of a nanotechnology consortium gives us easier access to address the general Constructive TA goal of interaction with nanoscientists and technologists; and within the ongoing developments of nanotechnology, we can make co-evolution more reflexive. We have gained credibility in the nanoworld, for example in European Networks of Excellence and in international conferences.

In nanotechnology, promises about new technological options abound, but little can be said definitively about their eventual realisation, let alone the impacts on society. As I have already noted, an impact assessment of nanotechnology will necessarily be speculative: one could call it science fiction (about future nanotechnology) combined with social science fiction (about the world in which future nanotechnology would have impacts). Still, it is important to try and anticipate and create controlled speculations about possible futures so as to stimulate reflection and broaden the scope of strategic choices about nanotechnology, and more generally.

While it is too early to expect actual conflicts about nanotechnology (even if demonstrations are occasionally staged), concerns are being articulated, for example about uncontrolled spread of nanoparticles. What happened (and continues to happen) with biotechnology and genomics is often seen as a lesson. Nanoscientists have actually called on governments to help avoid the impasse that has occurred with (green) biotechnology. (An example is Vicki Colvin, in the Hearings of the US Congress concerning the new Nanotechnology Bill, April 2003; see my discussion in Rip, 2006b). Genomics stimulation programmes have an ELSI component: Ethical, Legal and Social Issues, modelled after the Human Genome Programme in the 1990s. ELSI is now taken up, in various ways by European and American R&D stimulation programmes for nanotechnology. To include TA studies in the NanoNed programme is a similar proactive response to possible societal concerns. In practice, this need not imply a willingness to interact: it could work as a division of moral labour, where the scientists can point to the social scientists and ethicists as reasons why they themselves do not have to think about societal and ethical aspects (as happened to some extent with ELSI in the Human Genomics Program). When there is interaction, it requires ELSI scholars to link up (though not identify) with the perspectives of the nano-enactors.

While the actual future cannot be predicted, the occurrence of socio-technical dynamics and emerging irreversibilities implies that there is an *endogenous future*.

Embedded in the present are preferred directions, which imply that a trajectory will be followed (Dosi, 1982). Even if paths are created while 'walking' (Garud and Karnøe, 2001), emerging paths can be mapped, and the dynamics of their emergence can be analysed (prospective technology analysis) – and the results can be fed back to actors. This, by itself, is not new. It overlaps with technological roadmapping, where one reasons forward to desired performances and functionalities, and identifies barriers to be overcome, and then reasons back to the efforts necessary to achieve them. In some cases, such activities have become institutionalised, for example in the semiconductor industry where an international consortium makes such roadmaps and uses them for strategic coordination. Predictions based on Moore's Law serve as a guideline for strategic decisions, and thus become self-fulfilling.

Socio-technical scenarios, in contrast to roadmapping, are open-ended. They branch out into the future, and are structured by prospective technology analysis. An advantage in terms of interaction and reflexivity is the way such scenarios link up with technology enactors. Actors always work with partial and diffuse versions of scenarios to orient themselves and others. A social science supported TA will improve the quality of the scenarios. A structural problem is that enactors tend to project a linear future, defined by their intentions, and use the projection as a roadmap – only to be corrected by circumstances. Mapping tools which force actors to consider the non-linearity of evolution, and accept the complexity, can be developed to make them more effective (if the actors are prepared to accept such social science based support).

The approach combines:

- mapping and analysing the ongoing dynamics of technological development, the actors and networks involved, and the further (and possibly conflictual) embedding of the technology in society, with particular attention paid to emerging patterns, including preferred technological paths and so-called dominant designs, which will shape further co-evolution
- identifying and articulating socio-technical scenarios about further developments, possible impacts, forks in development and the possible choices of various actors. This stimulates technologists and other relevant actors to reflect on their strategies and choices, making these more socially robust
- organising interactive workshops and other 'bridging events' where a (relevant) variety of actors participate (up to critical NGOs), but always including nanoscientists and technologists. Socio-technical scenarios allow them to probe each other's worlds in a structured way.

Not only is nanotechnology at an early stage, it is an enabling technology: the promises of nanotechnology have to be realised through its uptake in other products and services (from faster DNA analysis to improved coatings, sunscreens and drug delivery). Its impact depends on what happens there, and is in that sense co-produced, even if one might consider that it is nanotechnology which has made the difference – thus, the well-known problem of attribution of impacts to earlier actions (Rip, 2001). In our recent work, we therefore also consider socio-technical scenarios for a sector, say food packaging, where nanotechnology might be taken up.

In the past the emphasis was on socio-*technical* scenarios. We are now developing *socio*-technical scenarios as well, e.g. for risks of nanoparticles, for food and nano. Societal issues, which in the socio-*technical* scenarios become visible only at a later stage and which reflect the concentric bias of enactors, and their implications for the governance of new and emerging science and technology will now be at centre stage: risks to health and environment, economic and distributional impacts, equity issues and sustainable development (particularly in developing countries), longer term and ethical issues.

Our approach to prospective analysis and assessment of a technology at an early stage can build on a number of relevant methodologies and pilot studies (including SocRobust, an EU-FP5 project, see Larédo et al., 2002)), as well on recent advances in disciplines such as innovation studies (itself an interdisciplinary domain), sociology of technology, evolutionary economics, industrial economics, industrial ecology and political science. Interestingly, the boundaries between these disciplines are fuzzy, and there is an increasing overlap. (..)

Building Scenarios and Modulating Views and Interactions
(Excerpt from Rip and Te Kulve (2008), pp. 51-53 and 66-67)

Emerging Irreversibilities and Endogenous Futures

While new (emerging) science and technology introduce novelties, and thus potentially breaking up existing orders to some extent, subsequent developments create new patterns, up to dominant designs and industry standards. In other words, irreversibilities emerge, which will be reinforced when actors invest in the paths that appear to emerge. "Emerging irreversibilities facilitate specific technological paths – make it easier to act and interact – and constrain others – make it more difficult to do something else" (Van Merkerk & Robinson 2006).

Emerging irreversibilities are a general feature of social life, and the sociological concept of 'institutionalization' captures a large part of what happens. When technology is involved, irreversibilities are further solidified in configurations that work (Rip & Kemp 1998). The concept of 'configuration that works' applies to artefacts and systems, and includes (in principle) social linkages and alignments as well. A dominant design or industry standard would be an example, where the actual dominance, and thus the 'working' of the design, depends on the adherence of relevant actors to it.

Paths and other stable patterns enabling and constraining actions and views will shape further development. Thus, they span up an "endogenous future": further developments are predicated on the pattern of the present situation. Not in a deterministic way: there are always choices and contingencies. Also, and important for the approach of Constructive TA, actors can use an understanding of these dynamics to act more productively, and in any case more reflexively.

The phenomenon as such of emerging and stabilizing socio-technical paths is now widely recognized. Actors anticipate them, up to attempts to create the "better" path, for example in the struggle about an industry standard or a dominant design. The battle over consumer videorecording in the 1970s and 1980s is an example (Cusumano et al. 1997; Deuten 2003), and is remembered and referred to, for example in the ongoing battle over advanced DVD standards.

The idea, and the practice, of roadmapping builds on the recognition of emerging paths, and the conviction that one can create such paths intentionally by coordinating actions. In micro-electronics there is a long history starting with SemaTech in the USA, and roadmapping is now globally coordinated by the International Semiconductor Technology Roadmap consortium. This is an example of a (strong) shaping of a path – as long as the strategic games based on mutual dependencies in the sector continue to be adhered to.

The idea of endogenous futures predicated on existing and emerging irreversibilities is also the starting point for our scenario exercises. Such scenarios reconstruct ongoing and future paths, their rise and fall, and how they become a reference for actors' strategies. Compared with roadmapping exercises, they are open ended: there is no future socio-technological functionality and performance that must be realized and thus become the starting point to identify challenges.

Modulation of Concentric Perspectives of Enactors

To address the aim of Constructive TA, to modulate and broaden technological innovation, it is necessary to understand our primary "target group", i.e. those actors directly or indirectly involved in developing new technology. The first step is to recognize that "enactors" will work within a concentric perspective. For example in the case of the development of new products, product managers often views the environment as concentric layers around the new product, starting with the business environment and ending with the wider society. Eventually, alignments with all layers need to be made, but the product manager often deals with them sequentially, starting first with clarifying functional aspects of the product, before addressing broader aspects (Deuten et al. 1997). The term 'enactor' is adapted from Garud and Ahlstrom (1997). Their analysis can be developed to create a theory of actors and interaction dynamics around new and emerging technologies.

Enactors, i.e. technology developers and promoters, who try to realize (enact) new technology, construct scenarios of progress, and identify obstacles to be overcome. They thus work and think in 'enactment cycles' which emphasize positive aspects. This includes a tendency to disqualify opposition as irrational or misguided, or following their own agendas. "How otherwise can one explain that progress is opposed?"[7] Enactors will get irritated, because for them, explaining the promise of their technological option should be enough to convince consumers/ citizens. For nanotechnology, enactors now also anticipate on obstacles similar to the ones that occurred for GMO (Genetically Modified Organisms) in agriculture and food, cf. Colvin (2003). But the structure of the situation remains the same, that of an enactment cycle.

While enactors identify with a technological option and products-to-be-developed, and see the world as waiting to receive this product – "the world" may well see alternatives, and takes a position of comparing and selecting. Thus, the other main position to be distinguished is the one of comparative selectors (not necessarily critics). There are professional comparative selectors (regulatory agencies like the US Food and Drug Administration) which use indicators, and develop calculations to compare the option with alternatives (e.g. versions of cost-benefit analysis). There are also citizens – consumers, etc., as amateur comparative selectors – which can range more freely because they are not tied to certain methods, and to accountability. Spokespersons for consumers, citizens react and oppose (rather than just select); some NGOs become enactors for an alternative (as when Greenpeace Germany pushed for better fridge – Greenfreeze).[8]

---

7   For an extreme example of this argument, see Bond (2005) who argues that it would be unethical to stop the development of nanotechnologies because of their potential to "enabling the blind to see and the deaf to hear."

8   Pressure to substitute fluorochlorocarbons as coolants were ineffective until Greenpeace Germany and an ailing refrigerator company in former East-Germany got together and created a technical alternative, Greenfreeze, which shifted the balance of forces, at least in Europe (Verheul & Vergragt 1995). Van de Poel (1998, 2003) has shown more generally that it is important to have a technological alternative, a configuration that actually works, to effect regime change.

Enactors can, and sometimes must, interact with comparative selectors. Formally as with the US Food and Drug Administration, or informally as in marketing and in the recent interest in interactions between strategic management of firms and spokespersons for environment and civil society. And in a "domesticated" version in test-labs like Philips Home-Lab (Philips Research – Technologies) and the RFID (Radio-Frequency Identification Device) -filled shop (RFID Journal 2003) in which people are invited to try out the new products, services and infrastructure.

The further step is to recognize that enactment cycles and comparative-selection cycles interfere anyway, and to identify (possible) interference locations and events *and what can happen there*. Garud & Ahlstrom (1997) speak of 'bridging events' and identify some examples and their limitations. Bridging events may not only include 'events', but also *structural* interaction. Cowan (1987)'s analysis in terms of a consumption junction is one example.

For the soft intervention approach of Constructive TA, an important modality is to support and orchestrate bridging events. This is creating and orchestrating spaces where interactions occur, even if the interactions between citizens/consumers and technology developers and promoters will always be partial (because of their difference in perspective). There will be "probing of each other's realities" (as Garud and Ahlstrom (1997) called it), with more or less contestation. In interactive workshops, this can be supported by socio-technical scenarios which show effects of (interfering) enactment and selection cycles, and give more substance to the interactions.

(…) (…) (…)

In Conclusion

We have seen that concentric socio-technical scenarios are appreciated by enactors. This derives from their own tendency to think in terms of scenarios (and opportunities and blockages). We expect that political and civil society actors in a comparative-selector position will appreciate the multi-level scenarios, because their role there is constitutive rather than contributory.

We have also argued that concentric scenarios need to be further contextualized, and include multi-level dynamics. If fully-fledged multi-level scenarios could be created, we would see how the present diagrams are actually selections, geared to the perspective of a particular kind of actor (in our diagrams, that of enactors). In other words, concentric scenarios are not just a ploy to accommodate enactors in order to broaden their views and actions – a necessary evil, as it were. Multi-level scenarios do not identify with a particular type of actor, but provide the backdrop to actor-specific scenarios. They are the scenarios related to CTA agents, who have no axes to grind other than promoting reflexivity (Schot and Rip 1997). They are broader, and in that sense better. But their broadness also makes them less relevant to actors, unless a translation and specification is made. Our suggestion that political and civil society actors would appreciate multi-level scenarios must therefore be modified: also for them, selection and specification is necessary.

The CTA approach combines analysis and action: from tracing dynamics and articulating them, to modulating co-evolution, at least making it more reflexive. Socio-technical scenarios are thus not just tools, supporting one or another type of actor in reflection and articulation of strategies. They are created (or co-created) by CTA agents as part of insertions in ongoing dynamics, unavoidably so. (..)

Further reflection on CTA as soft intervention is in order. If we are right in our diagnosis of endogenous futures, and use it to create socio-technical scenarios, we are actually creating a paradoxical situation, where we say to actors that they are part of a pattern and being shaped by it (cf. paths) – and then enjoin them to take action, perhaps changing the pattern. In each concrete case, actors may recognize how their choices and actions are being shaped (softly determined) by socio-technical factors and patterns, while at the same time they will act, and attempt to act better based on their understanding of such factors and patterns – up to undermining them.

This point about actors being part of a pattern that is reproduced, and then profiting from insight in the pattern to do something different occurs explicitly as soon as there are stabilized anticipations. The well-known Gartner Group hype-disappointment cycle (mainly applied to information and communication technologies) is a case in point.[9] Including its "paradoxical" use: there is an existing pattern (up to master curves), and The Gartner Group is willing to advise firm X about when and how to follow the cycle, or step out. Determinism and voluntarism in one: Things *will* go this way, but if you understand it (and hire Gartner Group as consultant) you can escape from it by acting. Similarly, one could say: emerging irreversibilities and path dependencies will occur, but if you understand them, thanks to Constructive TA, you can escape them …

Such paradoxes have to be kept in mind, but socio-technical scenarios and scenario workshops can do useful things. They contribute to reflexive co-evolution of science, technology and society. This need not, by itself, lead to a better technology in a better society. But it will definitely make the co-evolutionary processes more reflexive, and create openings for responsible innovation.

---

9   The hype-disappointment cycle is a folk-theory, because widely recognised, used to draw out implications, and not object of systematic research. It is a relatively innocent folk theory, though, because actors can easily recognize its limitations, and define their actions taking the limitations into account. See further Rip (2006b).

Futures of CTA

While the excerpts started with macro-level developments, the focus was on meso- and micro-level approaches and experiences. There is recognition, definitely in the world of enactors of nanotechnology, of the value of Constructive TA exercises, and one can argue that they are here to stay, particularly if the present discourse of responsible development of technologies becomes operationalized and create a demand for CTA approaches.

If the CTA approach becomes internalized in the world of technology (cf. Rip 2010, reproduced as Chapter 4), there is no need for CTA agents anymore – other than as consultants to be hired to do part of the job. This is one possible scenario for the embedding of CTA (as a "social technology") in society. Other scenarios where CTA activities become a further administrative burden for enactors of technology, and/or be disappointing because there appears to be little or no effect, have been considered by Te Kulve (2011) and Shelley-Egan (2011).

Reflection on the future of CTA must locate such approaches also in broader and long-term developments. And consider the goals of CTA: more reflexive co-evolution, and modulation to have better technology in a better society. One consideration is how far reflexivity and articulation of responsibilities, can be delegated to social scientists and humanities scholars doing "accompanying" research (Rip 2009). There are political dimension as well, for example how the future of CTA will be shaped by the tensions between neo-liberal ideologies & practices, and communitarian ideals & practices. Responsible development has a different complexion in these two ideologies. CTA, when practised more widely, may itself have societal effects because it will support neo-corporatist practices of societal ordering (decision making and implementation). There may then be concerns about loss of representative democracy. If the goal of a better technology in a better society is put upfront, a dose of neo-corporatism might actually be a good thing. My concern would not be about democracy, but about ensuring that there will opportunities to open up what tends to stabilize and close down (Rip 2006a).

The long-term goal of CTA to contribute to a better technology in a better society is ambitious, and it suffers from a structural ambivalence. There is the recognition of non-linearity (and contingency) in technological development and embedding in society. Still, CTA wants to intervene/modulate at an early stage, with the expectation that things will go better in later stages, somehow. Thus, paradoxically, the return of linearity. The recent phrase "benign-by-design" captures this ambivalence. It replaces the vanity of progress through technology by the reflexive vanity that we can overcome non-linearity because we have good intentions.

Vanities motivate people, at the risk of them neglecting side-effects of their actions. Modesty (about what one can achieve) and humility (in the face of complexity) are important to counteract the vanities that go with action in the real world. But we should not give up on such action. To that extent, I conclude by quoting myself quoting William the Silent:

> William the Silent, Prince of Orange, in the 16th century (supposedly) said: "Point n'est besoin d'espérer pour entreprendre, ni de réussir pour persévérer". This aphorism about being prepared to persevere even if there is no hope to realize one's goals, is widely quoted. And if acted upon, would imply that irony need not stifle action. This would save non-modernism from evaporating into reflexivity. (Rip 2006a)

Integrated Bibliography

Blume, S. (2000) Land of hope and glory: Exploring cochlear implantation in the Netherlands. *Sci Tech Hum Val, 25*, 139–166.

Bond, P. J. (2005). Responsible nanotechnology development. In Swiss Re Centre for Global Dialogue (Ed.), *Nanotechnology: Small size – large impact?* (pp. 7-8). Zurich: Swiss Reinsurance Company.

Collingridge, D. (1980). *The social control of technology*. London: Frances Pinter.

Colvin, V. L. (2003). *Testimony of Dr. Vicki L. Colvin, director Center for Biological and Environmental Nanotechnology (CBEN) and associate professor of Chemistry Rice University, Houston, TX before the US House of Representatives Committee on Science in regard to 'Nanotechnology Research and Development Act of 2003'*. Retrieved from http://www.house.gov/science/hearings/full03/apr09/colvin.htm

Cowan, R. S. (1987). The consumption junction: A proposal for research strategies in the sociology of technology. In W. E. Bijker, T. P. Hughes & T. Pinch (Eds.), *The social construction of technological systems* (pp. 261-280). Cambridge, Massachusetts and London, England: The MIT Press.

Cusumano, M. A., Mylonadis, Y., & Rosenbloom, R. (1997). Strategic manoeuvring and mass-market dynamics: The triumph of VHS over Beta. In M. Tushman & P. Anderson (Eds.), *Managing strategic innovation and change* (pp. 75-98). Oxford: Oxford University Press.

Daey Ouwens, C., Hoogstraten, P. van, Jelsma, J., Prakke, F., & Rip, A. (1987). *Constructief technologisch Aspectenonderzoek. Een Verkenning* (NOTA Voorstudie 4). Den Haag: Staatsuitgeverij.

Deuten, J. J. (2003). *Cosmopolitanizing technologies. A study of four emerging technological regimes*. Enschede: Twente University Press.

Deuten, J. J., & Rip, A. (2000). Narrative infrastructure in product creation processes. *Organization, 7*(1), 67–91.

Deuten, J. J., Rip, A., & Jelsma J. (1997) Societal embedment and product creation management. *Tech Anal Strat Manag, 9*(2), 219–236.

Dosi, G. (1982) Technological paradigms and technological trajectories: A suggested interpretation of the determinants and directions of technical change. *Res Pol, 6*, 147–162.

Garud, R., & Ahlstrom, D. (1997). Technology assessment: A socio-cognitive perspective. *J Eng Tech Manag, 14*, 25–48.

Garud, R., & Karnoe P. (2001). *Path dependence and creation*. Mahwah, N.J.: Lawrence Erlbaum Associates.

Geels, F. (2002). *Understanding the dynamics of technological transitions. A co-evolutionary and socio-technical analysis*. Enschede: University of Twente.

Grin, J., & van de Graaf, H. (1996). Technology assessment as learning. *Sci, Tech Hum Val, 21*, 72–99.

Grin, J., van de Graaf, H., & Hoppe, R. (1997). *Technology assessment through interaction. A guide*. Den Haag: Rathenau Institute.

Guston, D. H., & Sarewitz, D. (2002). Real-time technology assessment. *Tech Soc, 24*, 93–109.

Hoogma, R. (2000). *Exploiting technological niches: Strategies for experimental introduction of electric vehicles*. Enschede: Twente University Press.

Hoogma, R., Kemp, R., Schot, J., & Truffer, B. (2002). *Experimenting for sustainable transport: The approach of strategic niche management*. London: Spon Press.

Kemp, R., Rip, A., & Schot, J. (2001). Constructing transition paths through the management of niches. In R. Garud & P. Karnoe (Eds.), *Path dependence and creation* (pp. 269-299). Mahwah, N.J.: Lawrence Erlbaum Associates.

Laredo, P., Jolivet, E., Shove, E., Raman, S., Rip, A., Moors, E., Poti, B., Schaeffer, G.-J., Penan, H., & Garcia, C. E. (2002). *SocRobust (Management tools and a management framework for assessing the potetial long-term S&T options to become embedded in society) Final Report*; Project SOE 1981126 of the TSER Programme of the European Commission. Paris: Armines.

Ministerie van Onderwijs en Wetenschappen. *Integratie van Wetenschap en Technologie in de Samenleving. Beleidsnota*. 's-Gravenhage: Tweede Kamer 1983-1984, *18 421*(1-2). (Policy Memorandum: Integration of Science and Technology in Society)

Parandian, A., Rip, A., & Te Kulve, H. (forthcoming). Dual dynamics of promises and waiting games around emerging nanotechnologies. Contributed to Special Issue of *Technology Analysis & Strategic Management*.

Reuzel, R. (2004). Interactive technology assessment of paediatric cochlear implantation. *Poiesis & Praxis, 2*, 119–137.

Rip, A. (1992). Between innovation and evaluation: Sociology of technology applied to technology policy and technology assessment. *RISESST, 2*, 39-68. (also published in Italian: Tra innovazione e valutatione. La sociologia applicata alla politica ed alla valutaziome della tecnologia. In L. Cannavo (Ed.), *StudisSociali della tecnologia. Metodologie integrate di valutazione* (pp. 63-105). Roma: Euroma/La Goliardica.

Rip, A. (1992). A quasi-evolutionary model of technological development and a cognitive approach to technology policy. *RISESST, 2*, 69-102.

Rip, A., Misa, T., & Schot, J. W. (Eds.). (1995). *Managing technology in society. The approach of constructive technology assessment*. London: Pinter Publishers.

Rip, A., & Kemp, R. (1998). Technological change. In S. Rayner & E. L. Malone (Eds.), *Human choice and climate change* (2nd ed.). (pp. 327–399). Columbus, Ohio: Battelle Press.

Rip, A. (2000). There's no turn like the empirical turn. In P. Kroes & A. Meijers (Eds.), *The empirical turn in the philosophy of technology* (pp. 3-17). Amsterdam etc.: JAI, an imprint of Elsevier Science.

Rip, A. (2001). Assessing the impacts of innovation: New developments in technology assessment. In OECD Proceedings (Ed.), *Social sciences and innovation* (pp. 197–213). Paris: OECD.

Rip, A., & Schot, J. W. (2002). Identifying *loci* for influencing the dynamics of technological development. In K. Sorensen & R. Williams (Eds.), *Shaping technology, guiding policy; Concepts, spaces and tools* (pp. 158-176). Cheltenham: Edward Elgar.

Rip, A. (2002). *Co-evolution of science, technology and society*. Expert review for the Bundesministerium Bildung and Forschung, Forderinitiative: Politik, Wissenschaft und Gesellschaft (Science Policy Studies). Managed by the Berlin-Brandenburgische Akademie der Wissenschaften. Enschede: University of Twente.

Rip, A. (2006a). A co-evolutionary approach to reflexive governance – and its ironies. In J.-P. Voss, D. Bauknecht & R. Kemp (Eds.), *Reflexive governance for sustainable development. Incorporating unintended feedback in societal problem-solving* (pp. 82–100). Cheltenham: Edward Elgar.

Rip, A. (2006b). Folk theories of nanotechnologists. *Sci Cult, 15*(4), 349–365.

Rip, A. (2008). Constructive technology assessment of nanoscience and –technologies. In G. Hirsch-Hadorn, H. Hoffmann-Riem, S. Biber-Klemm, W. Grossenbacher-Mansuy, D. Joye, C. Pohl ... & E. Zemp (Eds.), *Handbook of transdisciplinary research* (pp. 145-157). Heidelberg: Springer.

Rip, A., & te Kulve, H. (2008). Constructive technology assessment and sociotechnical scenarios. In E. Fisher, C. Selin, & J. M. Wetmore (Eds.), *The yearbook of nanotechnology in society, Volume I: Presenting Futures* (pp. 49-70). Berlin etc: Springer.

Rip, A. (2009). Futures of ELSA. *EMBO Reports, 10*(7), 666-670.

Rip, A. (2010). De facto governance of nanotechnologies. In M. Goodwin, B.-J. Koops & R. Leenes (Eds.), *Dimensions of technology regulation* (pp. 285-308). Nijmegen: Wolf Legal Publishers.

Robinson, D. K. R. (2009). Co-evolutionary scenarios: An application to prospecting futures of the responsible development of nanotechnology. *Technological Forecasting & Social Change, 76*, 1222-1239.

Robinson, D. K. R. (2010). *Constructive technology assessment of emerging nanotechnologies. Experiments in interactions.* Enschede: University of Twente.

Ruivenkamp, M. (2011). *Circulating images of nanotechnology.* Enschede: University of Twente.

Schot, J. W. (1991). Technology dynamics: An inventory of policy implications for constructive technology assessment. In J. W. Schott (Ed.), *Maatschappelijke Sturing van technische Ontwikkeling. Constructief technology assessment als hedendaags Luddisme.* Enschede: University of Twente.

Schot, J., & Rip, A. (1997). The past and future of constructive technology assessment. *Technol Forecast Soc Change, 54*, 251–268.

Smith, M. R., & Marx, L. (Eds.). (1994). *Does technology drive history? The dilemma of technological determinism.* Cambridge, MA: MIT Press.

Te Kulve, Ha. (2011). *Anticipatory interventions in the co-evolution of nanotechnology and society.* Enschede: University of Twente.

Van den Belt, H., & Rip, A. (1987). The Nelson-Winter/Dosi model and synthetic dye chemistry. In W. E. Bijker, T. P. Hughes & T. J. Pinch (Eds.), *The social construction of technological systems. New directions in the sociology and history of technology* (pp. 135-158). Cambridge, Mass.: MIT Press.

Van de Poel, I. (1998). Changing technologies: A comparative study of eight processes of transformation of technological regimes. Enschede: Twente University Press.

Van de Poel, I. (2003). The transformation of technological regimes. *Research Policy, 32*, 49-68.

Van de Ven, A. H., Polley, D. E., Garud, R., & Venkatamaran, S. (1999). *The innovation journey*. New York, NY.: Oxford University Press.

Van Lente, H. (1993). *Promising technology - The dynamics of expectations in technological developments*. Enschede: University of Twente.

Van Merkerk, R. O., & Robinson, D. K. R. (2006). The interaction between expectations, networks and emerging paths: A framework and an application to lab on a chip technology for medical and pharmaceutical applications. *Technology Analysis and Strategic Management, 18*(3-4), 411-428.

Van Merkerk, R. (2007). *Intervening in emerging nanotechnologies. A CTA of Lab-on-a-Chip technology*. Utrecht: University of Utrecht.

Verganti, R. (1999). Planned flexibility: Linking anticipation and reaction in product development projects. *J Prod Innovat Manag, 16*, 363–376.

Verheul, H., & Vergragt, P. J. (2005). Social experiments in the development of environmental technology: A bottom-up perspective. *Technology Analysis & Strategic Management, 7*(3), 315-326.

Violino, B. (2003, April 27). Metro opens 'Store of the future'. RFID Journal. Retrieved from http://www.rfidjournal.com/articles/view?399

Chapter 6
The Past and Future of RRI*

Background

Within the space of a few years, the idea of Responsible Research and Innovation, and its acronym RRI, catapulted from an obscure phrase to an issue in the European Commission's Horizon 2020 Program, to a topic of conferences (Danish EU Presidency 2012) and edited volumes (Owen et al. 2013a). Without using the exact term, a discourse on responsible development of nanotechnology was visible already in the mid 2000s, and nanotechnology can be seen as the lead domain for discourse and activities on RRI.[1]

It is important to consider the nature of this phenomenon, of the rise of RRI, because it appears to mobilise various actors and will already for that reason have effects (Fisher and Rip 2013). Our understanding will necessarily be partial, because we are in the midst of the process. Still, one can make an attempt, even if it will necessarily be somewhat essayistic. There have been studies and reports on what might be the substance of RRI and its potential implementation; a good overview is given in Owen et al. (2013a). I want to raise more distantiated sociological questions: what are the roots of this phenomenon? What are the dynamics at present, and what do these imply for the future of RRI as a discourse, and as an emerging patchwork of practices? Such questions are now being raised, for example in the volume edited by Owen et al. (2013a) and in a conference in Oslo, December 2012.[2]

---

*   Source: *Life Sciences, Society and Policy* 2014, 10: 17 (electronic journal). DOI: 10.1186/s40504-014-0017-4

1   See Barben et al. (2007) and Rip (2010).

2   This was a dialogue conference, 4–5 December 2012, organized by the Oslo Research Group on Responsible Innovation, HiOA, located at the Oslo and Akershus University College for Applied Sciences, in cooperation with the Research Council of Norway's ELSA programme. For more details, see Forsberg (2014).

© Springer Fachmedien Wiesbaden GmbH, part of Springer Nature 2018   115
A. Rip, *Futures of Science and Technology in Society*, Technikzukünfte,
Wissenschaft und Gesellschaft / Futures of Technology, Science and
Society, https://doi.org/10.1007/978-3-658-21754-9_7

Papers deriving from that conference have been published in earlier issues of this journal, and are brought together in a Thematic Series entitled ELSA and RRI; this paper is the last in the series.[3]

My approach is to consider RRI as a social innovation that is gradually being articulated. What is being innovated are the roles and responsibilities of actors and stakeholders in research and innovation. Following Shelley-Egan (2011) and Rip and Shelley-Egan (2010), I will analyse this as a division of moral labour (an element in the overall cultural and institutional division of labour in societies), and position RRI in a historically evolving division of moral labour. This will then help me to trace the emerging path of RRI as a social innovation, and evaluate some of its features.

The historical-sociological approach is important to avoid limiting ourselves to a purely ethical perspective. I will introduce it briefly by comparing an earlier (16th century) issue of responsibility of scientists with a recent case which shows similar features. Broader responsibilities of scientists have been on the agenda, definitely after the Second World War and the shock (in the sense of lost innocence of physicists) of the atom bomb and its being used.[4] Thus, there is a past to RRI, before there was the acronym that pulled some things together. I say "some things" because there is no clear boundary to issues of responsibility linked to science. As a sociologist, I think of it as an ongoing patchwork with some patterns but no overall structure, where a temporary coherence and thrust can be created, now with the label RRI, which may then diverge again because patchwork dynamics reassert themselves.

With the benefit of the extended analysis of divisions of moral labour, informed by the notion of a language of responsibility, I can address the emerging path of RRI, including the reductions that occur, inevitably. These reductions, and in-stitutionalisation in general, are the reason to include some evaluation of future directions, and relate them to wider issues in the final comments.

---

3   Earlier papers in LSSP, part of the Series on ELSA and RRI, are Myskja et al. (2014), Zwart et al. (2014), Oftedal (2014) and Forsberg (2014).

4   As one response, the Bulletin of the Atomic Scientists was established in 1945. Kevles (1978): 335 quotes a leading physicist (James Franck) at the time as saying that scientists could "no longer disclaim direct responsibility for the use to which mankind ... put their disinterested discoveries." The development and use of the atom bomb was considered a watershed for mankind, particularly by German philosophers like Karl Jaspers and Günther Anders (see Van Dijk 1992).

An Evolving Division of Moral Labour

Let me start with a historical case, and compare it with a recent one in which similar features are visible. The 16th century Italian mathematician and engineer Tartaglia had to make a difficult decision, whether he would make his ballistic equation (to be applied to predict the trajectory of a cannon ball) public or not.[5]

In 1531 the Italian mathematician Nicola Tartaglia developed, inspired by discussions with a cannoneer from Verona whom he had befriended, a theory about the relation between the angle of the shot and where the cannon would come down. He thought of publishing the theory, but reconsidered: "The perfection of an art that hurts our brethren, and brings about the collapse of humanity, in particular Christians, in the wars they fight against each other, is not acceptable to God and to society." So he burned his papers (he had told his assistant Cardano about his theory, and Cardano published it a few years later).

But he changed his position, as he described it in his 1538 book Nova Scientia. "The situation has changed, with the Turks threatening Vienna and also Northern Italy, and our princes and pastors joining in a common defence. I should not keep these insights hidden anymore, but communicate them to all Christians so that they can better defend themselves and attack the enemy.

Now move forward to a case from 2013. In the online version of the *Journal of Infectious Diseases*, October 7, Barash and Arnon published their finding of the sequence of a newly discovered protein, but without divulging the actual sequence. The news item about this in *The Scientist Magazine* of 18 October 2013 says:

[This] represents the first time that a DNA sequence has been omitted from such a paper. "Because no antitoxins as yet have been developed to counteract the novel C. Botulinum toxin," wrote editors at The Journal of Infectious Diseases, "the authors had detailed consultations with representatives from numerous appropriate US government agencies."

These agencies, which included the Centers for Disease Control and Prevention and the Department of Homeland Security, approved publication of the papers as long as the gene sequence that codes for the new protein was left out. According to

---

5   I base myself here on a Dutch text, Bos (1975), who refers to Charbonnier (1928) for the story. Since I follow his text quite closely, I have used indents, even if it is not a quote in the strict sense.

New Scientist, the sequence will be published as soon as antibodies are identified that effectively combat the toxin, which appears to be part of a whole new branch on the protein's family tree.

There are other cases where possible publication of sensitive details are prohibited, by the US National Science Advisory Board for Biosecurity, as in the case of the bird flu research by the Rotterdam team led by Fouchier (see also Evans and Valdivia, 2012).

My point here is about the similarities of the two cases, including the trope of powerful knowledge (at least, that is how the scientists and others see it), and how it can be used and misused. In the cases, the primary response to the possibility of misuse was to keep this knowledge hidden, but this will depend on the situation and the evolving balance of interests and visions. Whether to make such knowledge publicly available, and in fact, whether to invest in developing it at all, has to be evaluated again and again.

Thus, the structure of the considerations is the same, but the difference is that in the 21st century, the decisions are not individual but part of formal and informal arrangements and authoritative decisions by advisory boards and government agencies. What is also interesting is that there is no reference to responsibility of the researcher/scientist. In the 16th century this was because the word did not yet exist. In the 21st century, it was because the focus is now on what is permissible and expected, rather than an own responsibility of the researchers. The division of moral labour has changed.

Before I continue to discuss present divisions of moral labour and how RRI can be positioned in that landscape, I need to briefly look at how the words 'responsible' and 'responsibility' have been used, and are still used, particularly to articulate roles and duties in an evolving social order, and then add how such roles can be part of long-term "settlements" of science in society (what is sometimes called a "social contract" between science and society, cf. Guston and Kenniston (1994)).

Elsewhere I have shown there is an evolving "language" of responsibility, in general and for scientists and scientific research (Rip 1981). The big dictionaries of modern languages (Oxford English Dictionary, Grande Larousse etc.) offer historical data on the use of words. The adjective (sometimes used as a noun, as in the French 'responsable') has been in use for a long time, in French since the 13th century, in English since the 17th century, but in a variety of meanings.[6] It is in the 18th century that stabilisation occurs into the pattern of meanings that we see nowadays.

---

6   The quotes in the Oxford English Dictionary suggest the meaning of 'responsible' was
     not stabilized, different authors could use it in their own way. "The Mouth large but not
     responsible (= correspondent) to so large a Body" (1698); "This is a difficult Question,

The noun "responsibility" is only used since the late 18th century: since 1782 in French, since 1787 in English (those are the earliest quotes presented in the dictionaries). It is important to keep the relatively recent emergence of the term "responsibility" in mind because the term is often used to refer to thoughts and analyses in texts of pre-19th century philosophers (e.g. Aristotle, Hume) who do not use the term. This suggests a continuity which is not there, and backgrounds societal developments through which the term "responsibility" emerged and obtained its meanings. But the sociogenesis of the concept of responsibility is not visible in handbooks and studies of morality in the past, because almost all authors tend to project present-day language usage onto the past.[7]

What happened at the turn of the 19th century and stabilized in the course of that century is the emergence of bourgeois society and the idea of citizens (*citoyens*) with their rights and duties. To articulate those, an extension of language was necessary – the language of responsibility. Through that language, it became possible to discuss and fill in social order concretely. And some outcomes would find a place in the formal Constitution of the nation states as they organized themselves. This language of responsibility remains important to discuss evolving social orders, in the small and in the large. And it has become important for scientists (the term itself being an early 19th century invention, see Ross 1962) and science.[8]

---

and yet by Astrologie responsible (= capable of being answered)". In the 17th century, the German language uses 'verantwortlich' in the sense of 'verantwort' (Grimm 1956), similarly Dutch 'verantwoordelijk' (Woordenboek der Nederlandse Taal 131), In German, this use continues, in Dutch it disappeared from regular use in the course of the 19th century (except for the use of 'onverantwoordelijk' in the sense of 'onverantwoord').

7   This tendency is frustrating in handbooks like the Dictionary of the History of Ideas (Wiener 1973) in which one would expect some sensitivity for historical developments. For example, in the Lemma on "free will and determinism" (vol. II, pp. 239–240) a brief sketch is offered of Hume's ideas, based on his Inquiry Concerning Human Understanding, Section VIII, using the terms "responsible" and "responsibility" all the time, while Hume himself speaks of "blameable" and "answerable" (and once of "accountable"). (Hume 1955, pp. 107–109). Somewhat of an exception is Adkins (1975) who limits the anachronism to his title, and emphasizes (in his introduction, p. 4) that moral responsibility is not an important concept for the Greeks (and does not occur as a term), because of their view of the world and society. It is only because of the Kantian turn, he claims, that a view of the world and society emerges in which "For any man brought up in a western democratic society the related concepts of duty and responsibility are the central concepts of ethics." (p. 2).

8   To avoid misunderstanding: I am not saying that this is the only meaning of responsibility. There is retrospective responsibility, visible in blaming and liability, and prospective responsibility, important because we are creating futures all the time (Rip 1981, Grinbaum & Grove 2013).

Before the notion of 'responsibility' had become important (and available at all) from the early 19th century onward, relations between science and society could already be at issue, in particular as a relatively protected space in exchange for acquiescence to the existing order (Rip 2011). In retrospect, one can see that a long-term "settlement" between science and society started in the late 17th century (in the 1660s in France and Britain, to be more precise), one indication being how the UK Royal Society was established with an implicit political charter: you can do science if you don't interfere in society.

The Business and Design of the Royal Society is: To improve the knowledge of natural things, and all useful Arts, Manufactures, Mechanick practices, Engynes and Inventions by Experiments - (not meddling with Divinity, Metaphysics, Moralls, Politicks, Grammer, Rhetorick, or Logick).[9]

It is clear that the Royal Society's founders avoided theology and politics; and in not meddling with „Grammar, Rhetorick, or Logick," the three basic disciplines of a university education, they also kept a distance between their „Business" and the universities.

This social contract between (emerging) science and society created a macro-protected space for science (Rip 2011), provided scientists showed prudential acquiescence to the powers that be.[10] Prudential acquiescence can actually be counteracted by a vision of progress through science which has served as a mandate for the autonomy of science, but could also lead scientists to become active in the wider world, as an embodied force of progress. This is quite visible in present newly emerging science and technology: scientists can speak for new and promising science (from astrophysics to cancer research) and for the importance of scientific approaches in improving the lot of mankind. Such messages can be taken up by others, and be further amplified (cf. below on narratives of praise and blame).

The overall settlement went through phases, with the ideal of an "ivory tower" coming into its own in the late 19th century[11], then broken open by claims of rele-

---

9   Robert Hooke's draft statutes (1663) of the Royal Society, quoted after Van den Daele (1978): 25. Van den Daele's overall analysis has informed (and inspired) my argument here.

10   The concept of 'prudential acquiescence' was introduced by Haberer (1969), p. 323, as a general feature of science. Rettig's (1971) point that there are exceptions is correct; however these are indeed exceptions. In other words, the macro-protected space not only protects, but also confines.

11   It could actually be applauded, as when a leading Dutch newspaper, *Het Nieuws van de Dag* (2 April 1908), referred to the world famous Dutch theoretical physicist J.D. van der

vance (already in and after the first World War), and contestation about relevance from late 1960s onward. The emergence of RRI might indicate that a next phase of the settlement is emerging, or at least that there are openings for it to happen.

Present Issues (Including RRI) in the Division of Moral Labour

There are justifications involved in claims and routines about roles and responsibilities, explicitly but also when they are taken for granted. A common view was captured in an aphorism by Ravetz (1975): "Scientists take credit for penicillin, but Society takes the blame for the Bomb". One can criticize such a view as gerrymandering, claiming the good and disavowing the bad (or at least, shifting the responsibility to others). But it is also a way of dividing labour (here, moral labour) to perhaps doing better overall. Think of the view that scientists have a moral obligation to work towards progress, and that is how they discharge their duty to society, while others (better qualified, or more responsible, or at risk) should look after social, ethical and political issues. What Ravetz's aphorism brings out is that one might want to query such a division of moral labour. This is a second-order ethics question: why would this be a "good" division of labour? Second-order ethics discusses the ethical (and more broadly, normative, cf. Rip 2013) aspects that become visible when one inquires into the justification of present overall social and institutional arrangements, rather than taking them for granted.

Developments since the 1970s, and particularly in the 2000s, are undermining this common justification and that is the reason why notions like responsible development of new technologies and responsible research and innovation have emerged. RRI need not go for implementation and good practices only, but can include second-order ethics and be a focus point to discuss the pros and cons of this division of moral labour.[12]

Waals, and asked rhetorically whether anyone would get a slice of bread more because of the Van der Waals equations. No, but that is exactly why we appreciate the cultivation of science (Rip and Boeker 1975: 458).

12 This need not be a one-sided critique of closed science. One consideration is that it is important to have the scientific endeavour be protected from undue interference. This is quite clear for the micro-protected spaces of laboratories and other sites of scientific work, and the meso-level protected spaces of scientific communities and peer review, although there is also opening-up, ranging from citizen science to criticism of scientific practices and the knowledge that is being produced (Rip 2011). Seen from the side of society, the scientific endeavour is legitimate as long as scientists deliver, both in terms

What RRI does already is to reinforce trends towards stakeholder and citizen engagement with science, and to some extent, innovation. It is useful to consider division of moral labour again, now for other actors than scientists.

Let me start with the well-known industrialist's argument about the need to go for profit to survive, while other actors should take care of second-order, possibly negative effects (most often, government actors are assumed to have this responsibility). While this argument continues to be heard, practices are different. The move towards corporate social responsibility is one example, and particularly important is the Responsible Care Program in the chemical sector (King and Lenox 2000). From a sociological perspective, one can see the importance of the notions of "good firms" and "cowboy firms" (or "rogue firms", cf. notion of "rogue states" on the global scene). The "good firms" behave well, according to a division of moral labour, and are to be praised for their efforts even if outcomes are not always ideal. While "cowboy firms" transgress and must be condemned, particularly because they endanger the credibility of the "good firms" in the sector.[13]

Analysis in terms of division of moral labour can also be used to understand the actual and possible role of lay people, citizens, and consumers. Consumers, for example, are projected as having a duty to buy, and be informed, and calculate rationally – if only to ensure that neo-classical economics remains applicable. But they can also go for political action through consumption decisions, including boycotts (cf. Throne-Holst 2012). And there are evolving liability regimes which shift the responsibilities between producers and consumers (cf. Lee and Petts (2013), particularly p. 153).

The present interest in public engagement often remains within traditional divisions of moral labour by positioning members of the public as articulating preferences which may then be taken up in decision making as additional strategic intelligence. But one could have joint inquiry into the issues that are at stake (Krabbenborg 2013). In Codes of Conduct (as for nanotechnology) and broader accountability of scientists and industrialists generally, there is an assumption that there will be civil society actors willing and able to call them into account. That may not be the case: civil society actors may not be able, or not be willing, to spend the necessary time and effort. This is already visible in so-called "engagement fatigue".

---

of their producing what is promised (progress, even if this can interpreted in different ways) and their adhering to the normative structure of science (cf. the issues of integrity of science). This is a mandate which justifies the relative autonomy of science – a sort of macro-protected space.

13 Interestingly, discussions about integrity of science and the occurrence of fraud have the same structure. Fraud is positioned as deviation from a general good practice, and done by "rogue scientists".

If one wants to overcome the traditional divisions of moral labour (for emancipatory reasons or because the present division of labour is not productive) other divisions of moral labour have to be envisaged and explored. One entrance point would be to consider evolving narratives of praise and blame (Swierstra and Rip 2007, Throne-Holst 2012) and turn them into blueprints of division of moral labour. This is a complex process, also because of the reference to possible future developments and the "shadow boxing" about the promises that ensues: Wonderful futures can be projected, waiting to be realised, which then justifies present efforts and allows criticism of those who don't want to join in.

Compare this quote from Philip J. Bond, US Under-Secretary of Commerce, 'Responsible nanotechnology development', in SwissRe workshop, Dec 2004:

> "Given nanotechnology's extraordinary economic and societal potential, it would be unethical, in my view, to attempt to halt scientific and technological progress in nanotechnology. (…) Given this fantastic potential, how can our attempt to harness nanotechnology's power at the earliest opportunity – to alleviate so many earthly ills – be anything other than ethical? Conversely, how can a choice to halt be anything other than unethical?"

What is not taken up in such sketches of a desirable world -- just around the corner, if only we would go forward without hesitation (in the quote, by pursuing nanotechnology) --, is the question of what makes these worlds desirable compared to other possibilities. It is a promise of progress, somehow, and when there is criticism, or just queries, rhetorics kick in. At the height of the recombinant DNA debate, second half of the 1970s, the medical possibilities were emphasized: "Each day we lose (because of a moratorium) means that thousands of people will die unnecessarily". The justificatory argument about GMO, in the contestation about its use in agriculture, now refers to hunger in developing countries (which need biotechnical fixes, it appears). If the promise is contested, a subsidiary argument kicks in: people don't understand the promise of the technology so we have to explain the wonders of the technology to them. (This is the equivalent of the well-known deficit model shaping exercises of public understanding of science.).

One sees here how narratives of praise and blame become short-circuited: only praise for the new technology is allowed. Some short-circuiting is inevitable, though. For example, hype about new technologies may be necessary to draw attention to them and mobilise resources for their further development – otherwise their promise would never materialise. Even while hype may lead to disappointment later on, as actors realise but apparently can't do much about (Rip 2006, Rip 2013). At the same time, there might be concerns about the nature and impacts of the new developments, which again can be exaggerated in order to get a hearing. There is shadow boxing all around. Still, there are implications for a division of moral labour. For example, the importance of early warning is now widely recognized (cf. Harremoës (2001) on late lessons of early warning), but who can tell, at an early stage, whether a warning is significant? Mandates might become articulated for who may legitimately warn. Critical NGOs might be candidates for the task of voicing concerns and pushing them on the agenda; but this can also be viewed as unnecessarily harassing the promoters of technology. RRI will have to confront these issues in actual practice, and may try to articulate rules, procedures (beloved by bureaucracies), and new "good" practices (the quotes are used to indicate that one cannot stipulate beforehand what will be "good").

Over time, there will be some solidification of divisions of moral labour, discursively, culturally, and institutionally. The emerging path of RRI as a social innovation is part of this broader institutionalization process. It will be shaped by it, but shape it as well, also by making it more reflexive.

A Path into the Future

Arguably, the cradle of RRI as a social innovation were the concerns about nanotechnology in the early 2000s. Concerns in two senses: concerns from commentators and NGOS about possible negative impacts of the new and still uncertain nanotechnology, and concerns from promoters that nanotechnology would face lack of acceptance and active resistance as had happened with biotechnology – so this time, they should do it right from the very beginning.[14] Part of "doing it right" was to communicate better with various publics. Another part was to talk of responsible development of nanotechnology, and explore possibilities what this might imply.

In the Mid-Term Review of the US National Nanotechnology Initiative, there was an attempt to loosely define responsible development of nanotechnology:

> Responsible development of nanotechnology can be characterized as the balancing of efforts to maximize the technology's positive contributions and minimize its negative consequences. Thus, responsible development involves an examination both of applications and of potential implications. It implies a commitment to develop

---

14 For the general observation, see Rip (2006). For the evocative phrase about doing it right from the very beginning, this summarizes the wording in Roco and Bainbridge (2001), p. 2, and was picked up on later, e.g. when presenting a risk framework for nanotechnology, developed in collaboration between the chemical firm Dupont and the USA NGO Environmental Defense Fund (Krupp and Holliday 2005).

and use technology to help meet the most pressing human and societal needs, while making every reasonable effort to anticipate and mitigate adverse implications or unintended consequences. (National Research Council 2006:73).

This attempt at definition reflects a consequentialist approach (and ethics) as in traditional technology assessment. This strand continues in the articulation of RRI, but is not the only strand, particularly in Europe where the interest is also in inclusive governance of new science and technology.[15]

There are more such calls for responsible development, especially of nanotechnology. An interesting example are the meetings of the International Dialogue on Responsible Research and Development of Nanotechnology, positioned as opening up a space for broad and informal interactions (Tomellini and Giordani 2008, see also Fischer and Rip 2013), but hopefully, having consequences. In the first meeting in 2004, there was a proposal to develop a Code of Conduct, which was eventually taken up by the European Union (see European Commission 2008). Interestingly, the Code is much broader than the consequentialist ethics visible in the review of the US National Nanotechnology Initiative; see in particular the reference to a culture of responsibility (N&N stands for Nanoscience and Nanotechnologies):

> Good governance of N&N research should take into account the need and desire of all stakeholders to be aware of the specific challenges and opportunities raised by N&N. A general culture of responsibility should be created in view of challenges and opportunities that may be raised in the future and that we cannot at present foresee (Section 4.1, first guideline).

Responsible development of nanotechnology, and the general idea of responsible innovation, have now become part of the policy discourse.[16] RRI is becoming an umbrella term, cf. the discussions leading to the European Commission's Horizon

---

15 'Inclusive governance' was an important goal for the European Commission since at least the early 2000s (European Commission 2003). It is not limited to new science and technology.

16 Stevienna de Saille (University of Sheffield), in her study of all documents pertaining to RRI (from the European Commission and others), concluded (personal communication) that the first occurrence of the term was in December 2007, to characterize the topic of a workshop with nanotechnologists and stakeholders, organized by Robinson and Rip 2007 (Robinson and Rip 2007). Robinson and I were picking up something that was in the air (while only half a year before, in an earlier attempt to organize such a workshop, we could not raise much interest among the members of the EU Network of Excellence Frontiers, our primary audience (Robinson 2010, p. 387–388)). We had not seen this term RRI used before, but thought of it to avoid a too narrow focus on risk issues in the workshop discussions. The later use of the phrase had other sources within

2020 Programme[17], while scientists already start to strategically use RRI in funding proposals (and are being pushed to do so by EU policy officers), and ethicists see opportunities to expand their business (even if they may have moral qualms about its implications).[18]

Branching out from responsible development of nanotechnology, and its precursor in the Human Genome Project's ELSI component, and ELSA studies more widely, there is now also consideration of responsible synthetic biology and geo-engineering, with or without reference to RRI.

Clearly, RRI is an attempt at social innovation, ranging from discursive and cultural innovation to institutional and practices innovation.[19] As with technological innovation, a social innovation is new and uncertain, and distributed. Because of the many and varied inputs, the eventual shape of the innovation will be a *de facto* pattern, with dedicated inputs. To get taken up, institutional changes and sub-cultural changes (where different actors have to change their practices) are necessary. Such changes can be stimulated by soft command and control, as when in the EU (and Member states) codes of conduct for RRI would be stipulated. But it is also a business proposition: to extend the 'social licence to operate' because of credibility pressures in/of society. And now also a link with working on so-called Grand Challenges (e.g. Owen et al. 2013b).

Responsible research and innovation implies changing roles for the various actors involved in science and technology development and their embedding in society. This is an important aspect of the social innovation of RRI, and reinforces

---

the European Commission. I mention our invention of the phrase mainly to pinpoint when it had become "in the air".

17 As EU Commissioner for Research, Innovation, and Science Máire Geoghegan-Quinn phrased it in her opening speech for the EU Presidency Conference on Science in Dialogue, towards a European model for responsible research and innovation, Odense, 23 April 2012: "*Horizon 2020 will support the six keys to responsible research and innovation...and will highlight responsible research and societal engagement throughout the programme*" (quoted from the official text handed out at the conference). Geoghegan-Quinn M. http://ec.europa.eu/commission_2010-2014/geoghegan-quinn/headlines/speeches/2012/documents/20120423-dialogue-conference-speech_en.pdf

18 The European Commission included, at the end of its 7th Framework Programme, Calls for background studies on RRI, to which ethicists, legal and governance scholars, and innovation studies scholars responded.

19 One innovative element is the shift in terminology, from responsibility (of individuals or organized actors) to responsible (of research, development and innovation). The terminology has implications: who (and where) lies the responsibility for RI being Responsible? This may lead to a shift from being responsible to "doing" responsible development.

its embedding in an evolving division of institutional and moral labour in handling new technology in society.[20] An example is how technology enactors cannot just delegate care about impacts to government agencies and societal actors anymore, while it is not clear yet what a new and productive division of labour and its specific arrangements might be.[21] Thus, RRI opens up existing divisions of moral labour, concretely as well as reflexively (the latter in the sense that it positions them as not necessarily given a priori, but constructed and constructable).

One can inquire into the kind of arrangements that should be created (*ab novo* and/or by modulating existing and evolving arrangements). There is a patchwork of considerations and experiments already. Strong trends are to organize new types of stakeholder interactions[22], and to have public engagement exercises as an input in development trajectories (Krabbenborg 2013).[23] Background considerations include the struggle between consequentialist ethics and virtue and good-life ethics. Dominant is the utilitarian ethics perspective: maximize technology's positive contributions and minimize negative consequences. And a neo-liberal version of it: it is enough if actors avoid causing harm. There is also the narrative of containment: keep hazards at bay, then there is no problem with a new technology, and enactors (developers and promoters who "enact" new technology, see Rip (2006) for the background to this terminology) can do what they want. The rise of ELSA studies and their equivalents functions as partial compensation for just pushing new ("promising") technology, and thus legitimizes continued development.

---

20 The earlier division of labour around technology is visible in how different government ministries and agencies are responsible for "promotion" and for "control" of technology in society (Rip et al. 1995). There is more bridging of the gap between "promotion" and "control", and the interactions open up possibilities for changes in the division of labour.

21 The reference to 'productive' is an open-ended normative point, a Kantian regulative idea as it were. It indicates that arrangements (up to the de facto constitution of our technology-imbued societies) may be inquired into as to their productivity, without necessarily specifying beforehand what constitutes 'productivity'. That will be articulated during the inquiry.

22 Cf. Constructive TA with its strategy-articulation workshops (Robinson 2010), where mutual accommodation of stakeholders (including civil society groups) about overall directions occurs – outside regular political decision-making.

23 In both cases, traditional representative democracy is sidelined. This may lead to reflection on how our society should organize itself to handle newly emerging technologies, with more democracy as one possibility. There have been proposals to consider technical democracy (Callon et al. 2009) and the suggestion that public and stakeholder engagement, when becoming institutionalized, introduce elements of neo-corporatism (Fisher and Rip 2013: 179).

These considerations and ongoing attempts will continue, and some will stabilize, contributing to an emerging path in the development of the social innovation of RRI. This has to do with entanglements, how actors refer to RRI and take it up in their repertoires, filling it in, and in particular ways. Inevitably, there will also be reductions, some productive, some less so, or perhaps just premature.[24]

In ongoing practices, whether these refer explicitly to RRI or not, we see reductions to create some tractability: a focus on upstream (to assure acceptance!?) – while the real challenges might be downstream. And a focus on risk issues – which appear to be more tractable than societal and ethical issues. These reductions can close down broader reflexivity, and definitely shape development, e.g. through evolving narratives of praise and blame. One example would the acceptance of versions of due process argument: "Was there upstream interaction with society? OK, enactor, then you cannot be blamed for what happens afterwards."

Evolving paths become settled in institutional arrangements, and the new arrangements are taken as given, somehow. While this is understandable, given the need to act and make a difference, and thus reduce complexities to concrete activities ("Let's get on, we're doing responsible development, we're OK"), it can be problematical, already because of the lock-ins (i.e. path dependencies) while society (and technology) change. One should ask "is this still a good arrangement?" This is not an argument for remaining open-ended in general, for its own sake, but to advocate monitoring the reductions that occur, and evaluating the directions that appear to be taken. There are no pre-given evaluation criteria, but the attempt, and the process, of such evaluations is important.[25]

---

24 In an earlier article in this series, Zwart et al. (2014) emphasize that in RRI, compared with ELSA, "economic valorisation is given more prominence", and see this as a reduction, and a reduction they are concerned about. However, their strong interpretation ("RRI is supposed to help research to move from bench to market, in order to create jobs, wealth and well-being.") appears to be based on their overall assessment of European Commission Programmes, rather than actual data about RRI. I would agree with Oftedal (2014), using the same references as he does, that the emphasis is on process approaches in which openness, transparency and dialogue are important.

25 With RRI becoming pervasive in the EU's Horizon 2020, and the attendant reductions of complexity, this is a concern, and something might be done about it in the sub-program SwafS (Science with and for Society). See http://ec.europa.eu/research/horizon2020/pdf/work-programmes/science_with_and_for_society_draft_work_programme.pdf

Final Comments

The advent and further development of RRI (as a social innovation) is part of larger processes, is being shaped by them, and in a small way, contributes to them. It is creating openings in existing divisions of moral labour, not just of scientists and technologists, but also industrialists, government actors and society actors. That is to be be valued, and while some closing down is necessary, that should occur reflexively.

How is the RRI path shaping up, and how will it settle? The role of the European Commission, as a catalyst in multi-level dynamics (Fisher and Rip 2013), will be important.[26] Scientists will continue to be prudentially acquiescent, but under the RRI regime they will now more often be held to account.[27] A consequence is that impact, or better, embedding in society, will be seen as part of professional responsibility of scientists, even if most often they cannot do much about it.

Industrialists are facing customers all the time. For newly emerging science and technology, these customers are often other businesses, while end users enter the picture only at one or two removes. With RRI, the responsibilities of industrialists are extended. One effect might be more interaction across the product-value chain.

NGOs and Civil Society Organisations come in as 'third parties', and are invited by the European Union and some government agencies to participate in responsible development of new technologies, even when they are not equipped (or willing) to do so (Krabbenborg 2013).

When these continue and stabilize, they add up to a master narrative for the further development of the social innovation of RRI. While there are explicit policy attempts (at least in the European Union) to create such an RRI path, what actually happens is an effect of ongoing struggles among many actors, and the possibility and desirability of such a path is one item in the struggles. Actually, one possible

---

26 The European Union's activities are more than creating funding opportunities, there can be effects in the longer term. The Framework Programmes, for example, have created spaces for interactions across disciplines and countries, and particularly also between academic science, public laboratories and industrial research, which are now generally accepted and productive. The emergence of these spaces has been traced in some detail for the programmes BRITE and ESPRIT in the early 1980s, by Kohler-Koch and Edler (1998).

27 This does not just derive from the RRI regime. Generally, the protected spaces for science are opening up. With the call for relevance since the 1970s, spokespersons (including sponsors of science) for possible clients have become involved, for example 'industry', or 'sustainability'. Another new development is that citizens are becoming active in knowledge production and even some quality control, with attendant deprofessionalization of scientific knowledge production.

development is that RRI as a label for a policy fashion loses its force. But even then, some good practices will have evolved and remain. An example would be the emerging interest in extended impact statements when submitting proposals to funders, and some competence how to do them, and how to assess them.

Clearly, the "settlement" of science in society is changing, with the new discourse of RRI and the related practices being one element, reflecting these changes as well as pushing them. Does this amount to a shift in how our societies order themselves, at least with respect to newly emerging science and technology, which is similar to the emergence of responsibility language and practices in the early 19th century? Not by itself, but it is part of a broader movement towards increasing social accountability of professionals and porousness of institutions which authors like Ulrich Beck have tried to capture with the notion of reflexive modernisation. While we need not follow Beck in his ideas about the necessary triumph of "second modernity" (Beck et al. 2003), the future of RRI is bound up with such larger changes, depending on them but also contributing to them.

Acknowledgements

The author is grateful to two anonymous reviewers who pointed out several weaknesses in the originally submitted version of the paper. The background perspective of the paper has benefitted from discussions with Pierre Delvenne (University of Liège).

References

Adkins A. W. H. (1975). Merit and responsibility. A study in Greek values. Chicago, IL: Chicago University Press.

Barash, J. R., & Arnon, S. S. (2014). A novel strain of clostridium botulinum that produces type B and type H botulinum toxins. *The Journal of Infectious Diseases, 209*(2), 183–191. https://doi.org/10.1093/infdis/jit449

Barben, D., Fisher, E., Selin, C., & Guston, D. H. (2007). Anticipatory governance of nanotechnology: Foresight, engagement, and integration. In E. J. Hackett, O. Amsterdamska, M. Lynch & J. Wajcman (Eds.), *The handbook of science and technology studies* (3rd. ed.). (pp. 979–1000). Cambridge, MA: MIT Press; 2007.

Beck, U., Bonβ, W., Lau, C. (2003). The theory of reflexive modernization: Problematic, hypotheses and research programme. *Theory, Culture & Society, 20*(2), 1–33. doi: 10.1177/0263276403020002001.

Bos, H. J. M. (1975). Mathematisering en maatschappij, of Hoe loopt een success-story af? Amsterdam: Mathematisch Instituut.

Callon, M., Lascoumes, P., Barthe, Y. (2009). Acting in an uncertain world: An essay on technical democracy. Cambridge, MA: MIT Press.

Charbonnier, P. (1928). Essais sur l'histoire de la ballistique. Paris: Societe d'Editions Geographiques, Maritimes et Coloniales.

Danish EU Presidency. (2012). *Conference on science in dialogue, towards a European model for responsible research and innovation.* Odense: 23-25 April.

European Commission. (2003). *Report from the Commission on European Governance.* Luxembourg: European Communities.

European Commission. (2008). *Commission recommendation of 07/02/2008 on a code of conduct for responsible nanosciences and nanotechnologie̦ research.* Brussels: 07/02/2008, C(2008) 424 final.

Evans, S. A. W., & Valdivia, W. D. (2012). Export controls and the tensions between academic freedom and national security. *Minerva, 50*(2), 169–190. https://doi.org/10.1007/s11024-012-9196-4

Fisher, E., & Rip, A. (2013). Responsible innovation: Multi-level dynamics and soft intervention practices. In R. Owen, J. Bessant, & M. Heintz (Eds.), *Responsible innovation* (pp. 165–183). Chichester, UK: John Wiley & Sons, Ltd. https://doi.org/10.1002/9781118551424.ch9

Forsberg, E.-M. (2014). Institutionalising ELSA in the moment of breakdown? *Life Sciences, Society and Policy, 10*(1), 1. https://doi.org/10.1186/2195-7819-10-1

Grande Larousse de la langue francaise. (1977). Paris: Librairie Larousse.

Grimm, A. (1956). Grimms Deutsches Worterbuch (bearbeitet von E. Wulcker, R. Merinzer, M. Leopold, C. Wesle und der Arbeitsstelle des Deutschen Worterbuches zu Berlin). Leipzig: Verlag von S. Hierzel.

Grinbaum, A., & Grove, C. (2013). What is "responsible" about responsible innovation? Understanding the ethical issues. In R. Owen, J. Bessant, & M. Heintz (Eds.), *Responsible Innovation. Managing the Responsible Emergence of Science and Innovation in Society* (pp. 119–142). Chichester: John Wiley & Sons.

Guston, D. H., Kenniston, K. (1994). The social contract for science. InD. H. Guston & K. Kenniston (Eds.), *The fragile contract. University science and the Federal Government* (pp. 1–41). Cambridge, MA: MIT Press.

Haberer, J.(1969). *Politics and the community of science.* New York, NY: Van Nostrand Reinhold.

Harremoes, P. (Ed.). (2001). *Late lessons from early warnings: The precautionary principle 1896-2000.* Copenhagen: European Environmental Agency.

Hume, D. (1955). An inquiry concerning human understanding. In W. H. Charles (Ed.). Indianapolis: The Liberal Arts Press (Bobbs-Merrill). Originally published in 1748.

Kevles, D. J. (1978). *The physicists. The history of a scientific community in modern America.* New York: Alfred A. Knopf.

King, A. A., & Lenox, M. J. (2000). Industry self-regulation without sanctions: The chemical industry's responsible care program. *Academy of Management Journal, 43*(4), 698–716. https://doi.org/10.2307/1556362

Kohler-Koch, B., & Edler, J. (1998). Ideendiskurs und Vergemeinschaftung: Erschließung transnationaler Räume durch europäisches Regieren. In B. Kohler-Koch (Ed.), *Regieren in entgrenzten Räumen, Politische Vierteljahresschrift, Sonderheft 29/1998* (pp. 169–206). Opladen/Wiesbaden: Westdeutscher Verlag.

Krabbenborg, L. (2013). *Involvement of civil society actors in nanotechnology: Creating productive spaces for interaction* (Unpublished doctoral dissertation). RU Groningen, Groningen.

Krupp, F., Holliday, C. (2005, June 14). Let's Get Nanotech Right. *Wall Street Journal,* p. B2.

Lee, R. G., & Petts, J. (2013). Adaptive governance for responsible innovation. In R. Owen, J. Bessant & M. Heintz (Eds.), *Responsible innovation* (pp. 143–164). Chichester: John Wiley & Sons.

Myskja, B. K., Nydal, R., & Myhr, A. I. (2014). We have never been ELSI researchers – there is no need for a post-ELSI shift. *Life Sciences, Society and Policy, 10*(1). https://doi.org/10.1186/s40504-014-0009-4

National Research Council. (2006). *A matter of size: Triennial review of the National Nanotechnology Initiative.* Washington, D.C.: National Academies Press.

Oftedal, G. (2014). The role of philosophy of science in Responsible Research and Innovation (RRI): the case of nanomedicine. *Life Sciences, Society and Policy, 10*(1). https://doi.org/10.1186/s40504-014-0005-8

Owen, R., Bessant, J., & Heintz, M. (Eds.). (2013). *Responsible innovation. Managing the responsible emergence of science and innovation in society.* Chichester: John Wiley & Sons.

Owen, R., Stilgoe, J., Macnaghten, P., Gorman, M., Fisher, E., & Guston, D. (2013). A framework for responsible innovation. In R. Owen, J. Bessant & M. Heintz (Eds.), *Responsible Innovation* (pp. 27–50). Chichester: John Wiley & Sons. Oxford English Dictionary.

Ravetz, J. … et augebitur Scientia (1975). In R. Harre (Ed.), *Problems of Scientific Revolution: Progress and Obstacles to Progress in the Sciences* (pp. 42–57). Oxford: Clarendon Press; 1975.

Rettig, R. (1971). Science, technology, and public policy. Some thematic concerns (review article). *World Politics,* (Jan), 273–293.

Rip, A. (1981). *Maatschappelijke verantwoordelijkheid van chemici.* Leiden: University of Leiden.

Rip, A. (2006). Folk Theories of Nanotechnologists. *Science as Culture, 15*(4), 349–365. https://doi.org/10.1080/09505430601022676

Rip, A. (2010). De facto governance of nanotechnologies. In M. Goodwin, B.-J. Koops & R. Leenes (Eds). *Dimensions of technology regulation* (pp. 285–308). Nijmegen: Wolf Legal Publishers.

Rip, A. (2011). Science in the context of application: Methodological change, conceptual transformation, cultural reorientation. In M. Carrier & A. Nordmann (Eds.), *Protected spaces of science: their emergence and further evolution in a changing world* (pp. 197–220). Dordrecht: Springer.

Rip, A. (2013). Pervasive normativity and emerging technologies. In S. van der Burg & T. Swierstra (Eds.), *Ethics on the laboratory floor* (pp. 191–212). Basingstoke: Palgrave Macmillan.

Rip, A., & Boeker, E. (1975). Scientists and social responsibility in the Netherlands. *Social Studies of Science, 5*(4), 457–484. https://doi.org/10.1177/030631277500500406

Rip, A, Misa, T. J., & Schot, J. (Eds). (1995). *Managing technology in society. The approach of constructive technology assessment.* London and New York: Pinter Publishers.

Rip A., & Shelley-Egan, C. (2010). Understanding public debate on nanotechnologies: options for framing public policies: A working document by the services of the European Commission. In R. Von Schomberg & S. Davies (Eds.), *Positions and responsibilities in the 'real' world of nanotechnology* (pp. 31–38). Brussels: European Commission.

Robinson, D. K. R. (2010). *Constructive technology assessment of emerging nanotechnologies. Experiments in interactions.* Twente: University of Twente.

Robinson, D. K. R., & Rip, A. (2007). *Responsible research and innovation. Towards a best nano practice* (Preparatory material for a workshop of the Frontiers Network of Excellence). Enschede: University of Twente.

Roco, M., & Bainbridge, W. S. (Eds.). (2001). *Societal implications of nanoscience and nanotechnology.* Boston, MA: Kluwer Academic Publishers.

Ross, S. (1962). *Scientist: The story of a word. Annals of Science, 18*(2), 65–85. https://doi.org/10.1080/00033796200202722

Shelley-Egan, C. (2011). *Ethics in practice: Responding to an evolving problematic situation of nanotechnology in society.* Enschede: University of Twente.

Swierstra, T., & Rip, A. (2007). Nano-ethics as NEST-ethics: Patterns of moral argumentation about new and emerging science and technology. *NanoEthics, 1*(1), 3–20. https://doi.org/10.1007/s11569-007-0005-8

Throne-Holst H. Consumers, Nanotechnology and Responsibilities. Operationalizing the Risk Society. Enschede: PhD thesis, University of Twente, defended 18 April 2012; 2012.

Tomellini, R., & Julien, G. (2008). *Report: Third international dialogue on responsible research and development of nanotechnology.* Brussels: European Commission.

Van den Daele, W. (1978). The ambivalent legitimacy of the pursuit of knowledge. In B. Egbert & G. Michael (Eds.), *Proceedings of the Conference Science, Society and Education* (pp. 23–61). Amsterdam: Free University Bookshop.

Van Dijk, P. (1992). Gunther Anders: de 'geantiqueerdheid' van de mens. In De Maat van de Techniek (Ed.), (pp. 98–138). Baarn: Ambo.

Wiener, P. P. (Ed.). (1973). *Dictionary of the history of ideas; Studies of selected pivotal ideas.* New York, NY: Charles Scribner's Sons.

Woordenboek der Nederlandse Taal . bewerkt door Dr. F. De Tollenare, m.m.v. A.J. Persin en J. Ph. Van Oostrom, deel XIX (1) 's Gravenhage en Leiden: Martinus Nijhoff en A.W. Sijthoff; 1959.

Zwart, H., Landeweerd, L., & van Rooij, A. (2014). Adapt or perish? Assessing the recent shift in the European research funding arena from 'ELSA' to 'RRI.' *Life Sciences, Society and Policy, 10*(1). https://doi.org/10.1186/s40504-014-0011-x

Chapter 7
Technology as Prospective Ontology*

Ontology

Engineers add to the furniture of the world, and thus shift its ontology – if we use the term "ontology" in a simplistic way (Rip 2000: 8). This 'adding" is not a simple, linear activity of first making something, and making it available, which is then added to the world. There is a strong prospective element. Artefacts start as technological options, a promise of functionalities, in other words 'hopeful monstrosities' (Mokyr 1990, Stoelhorst 1997). This is visible, sometimes literally, in the prototypes: these embody a prospective. When they are developed further, introduced and taken up on location, they remain unfinished. Technologies are configurations that work (Rip and Kemp 1998), but always precariously. In a sense, in their practices technologies are unruly (Wynne 1988).

Scenarios, embedded in the configurations, are an integral part of their working, including the prospect of a world in which they can function optimally, at first as a "fictive script" (De Laat 1996, 2000). A key element of such a script is that the promise of a technological configuration can be realised only by changing the world so that it can accommodate the new technological options. Artificial fertilisers were effective only if the land they were applied to was reshaped so as to resemble the test plots, i.e. the circumstances under which they had been tried out. Atomic energy required a reorganisation of liabilities (insurance companies did not want to carry all of the risk) and extensive safety measures.

In general, the configurations constructed by engineers promise functionalities, but it takes time and effort to realize them (precariously). Also because the world has to adapt to their "fictive script", and need not accommodate fully. In other words,

---

*   Source: *Synthese.* 168 (2009) 405-422.

they are always prospective configurations (cf. Van Lente & Rip 1998 on prospective structures to be filled in by agency), and remain so because they are never finished.

Why use the term "ontology"? The notion of ontology as "furniture of the world" is not very sophisticated philosophically, but it serves to introduce the topic of my article.[1] It can be read as similar to the pre-Socratic idea of the "stuff" of the world, with the additional connotation that the shapes of the "stuff" will evolve. This "stuff" of the world has a prospective element, not because there are promises and "fictive scripts" being made, but because the future is already there, prefigured in the present and evolving configuration.[2]

The idea of prospective ontology actually implies a general ontological point, even if I will develop it mostly in terms of technology. This general point derives from my work in actor-network theory, even if it does not depend on the details of this theory. In Callon et al. (1986) a distinction was made between existing & evolving actor-networks, i.e. assemblages of circulating "intermediaries" which add up to actors (human and non-human) but which can be decomposed again into the networks out of which they are built, and actor-worlds, the projections of future worlds, like "fictive scripts" (De Laat 1996, 2000). At any one moment, the evolving "stuff" of the world is a patchwork of actor-networks. The evolution is shaped by the actor-worlds contained in them and sometimes articulated, and the responses to them in terms of circulating "intermediaries.

---

1   The concept of 'ontology' as used in information science specifies the units or elements that will make up the software world, often with the additional requirement that these units resemble the units in the real world that is to be modelled in the software. In contrast, the notion of ontology I am using here is open-ended: it need not and cannot be fully specified, its "units" will be discovered and articulated in practices. The debate on the reduction of a chemical ontology to a physics ontology or its autonomy sits in between, because it can be limited to the epistemological status of entities like molecules, as perhaps just a specification of the kind of "software" that the discipline of chemistry will use (cf. Lombardi and Labarca 2005). Such usages of the term "ontology'" are widespread. Nersessian (2006: 131) discusses the "ontology "of artefacts in a lab as its furniture, with devices, instruments and equipment. Van de Ven and Poole (2005) discuss alternative "ontological views of organizations as things and organizing as processes" (p. 1377), and quote Tsoukas (2005) on two versions of the social world: "one, a world made of things on which processes represent change in things; the other, a world of processes in which things are reifications of processes" (p. 1379). They do refer, following Rescher (1996), to a philosophical tradition of process ontologies, including Whitehead's notion of ongoing activities "prehending" what goes on in their environment (p. 1378). Later, they take this up again as "temporal predispositions" (p. 1391), similar to what I will call embedded anticipation.

2   Cf. Dupuy and Grinbaum (2006) at p. 312, about the "common but mistaken conception of the future as unreal".

At the time, we experimented with representations as in Figure 1, a timeline of actor-networks and actor-worlds.

Fig. 1

Time-line of actor-
networks (AN) and
actor-worlds (AW)

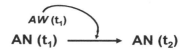

$$AW(t_1)$$
$$AN(t_1) \longrightarrow AN(t_2)$$

Such a representation can be read as how strategy and planning are commonly visualized: anticipation and then feedback into action. The point, however, is that anticipations and networks are an evolving whole, changing actor-networks are *de facto* enactment of overlapping and contrasting actor-worlds. Still keeping close to the strategy and governance literature, I have taken this up as "anticipation-in-action" (Rip 2006). For ontology, it implies a monistic view: the future is part of the ontology, not separate as just human projection.[3] This can be brought out by emphasizing that expectations are embodied, and conversely, that the material has a narrative character (sections 2 and 3).[4]

This is very visible in technology, and that is why a discussion of technology allows me to make the general ontological point. I shall use nanotechnology as a recurrent example, because it is not just a promising technology (up to projections of a third industrial revolution), but also requires us to re-think, and re-do, exper-

---

3 The monism is a monism of process ontologies, e.g. Whitehead (1929) on the actual world as process, and the process is a becoming of actual entities (see) and in another vein Bergson (1911) – on the flow of the real – , and his contemporary successor Deleuze (as Wood 2002, p. 157 phrases it). While both Whitehead and Bergson experience a revival (cf. Barry (2001, pp. 154-155), and Callon (1999)), I want to avoid their reliance on creativity, and the idea (mainly of Whitehead) that 'experience' pervades everything. Instead, I emphasize the prospective, and take technology (and landscape) as my entrance point rather than the living. In this paper, my intellectual strategy is to start (also in later sections) with commonsensical discussions of technology – and simplistic ideas of ontology – and then address further and deeper questions.

4 Conversely, narratives are material, even when they are taken up as just a story. They are voiced on location, or embodied/embedded in ink on pages of text which are lay-outed, bound in a book or some other concrete package.

iment and seeing – which at the nano-level is better thought of as indirect feeling (Baird et al. 2004). So it is an opportunity to consider ontology.

Having said this, I add, to avoid misunderstanding, that technology should not be limited to control and intentional assembling of building blocks. Also, when I say "prospective", I do not just think of design, as in designing an artefact or a system, which would be like the strategic planning reading of Figure 1: design first on paper, then assemble the prototype and tinker with it (the latter would already be design-in-action). I want to get away from the strong imagery of "building blocks", which is very common in mechanical engineering and is sometimes extended to technology in general. "Building the world atom by atom" is one of the slogans used for the promise of nanotechnology in the USA (from IWGN 1999 onward). Taken literally, it is nonsensical, but the phrase is thought to convey the message about the possibilities of the new technology to "build" whatever we want. While in fact, design is always open-ended, and in particular in nanotechnology, where the best one can do is to manipulate by trial-and-error and select what turns out to work,[5] or to induce self-organisation of materials, e.g. of thin layers on a surface.

Complementary to the imagery of "building blocks" is the idea of control, an idea which should be deconstructed. The competence of engineers is defined as one of control, and such control is predicated on their knowledge of the building blocks of reality.[6] But that knowledge is always partial, and will evolve through their interactions with the world.[7] Through such interactions, however, reality is changed, intentionally and unintentionally. In other words, advance control of eventual outcomes is impossible, because the work to realize them changes the validity of the knowledge on which the ideal of control was based. In fact, repairing if things work out differently is a key component of the competence of engineers.

The limited extent of advance control is even more striking when induced self-organisation is the way to realize an artefact. Synthetic chemistry relies on partially understood self-organization, and can be seen as exemplifying another scientific/engineering tradition than physics and mechanical engineering. The two traditions are now competing in nanotechnology (Bensaude-Vincent 2006).

In the chemical tradition, the "stuff" of the world is the starting point (Stein 2004). Nordmann's reading of Heidegger can be added here: "Heidegger (1977) offers

---

5   This, by the way, is also how semiconductor manufacture started out in the 1950s.

6   This is a suggestive turn of phrase. Mol (1999), discussing technology and politics, notes: "… along with this it was assumed that the building blocks of reality were permanent: they could be uncovered by sound scientific investigation."

7   Compare Schön (1983)'s notion of "conversation with the situation" – where the situation will talk back.

an account according to which technical control presupposes a causal picture of the world, one in which actions either poetically bring forth what lies dormant or instrumentally exploit a scheme of means-end relations." (Nordmann 2006, p. 52). My point is that "poetically bringing forth" effects is what characterizes technology, but without necessarily a determinate causal picture being involved.

A further aspect is delegation to realize effects, but not to building blocks that obey the wishes of their self-anointed master. Technology is open-ended, unruly, and in that sense out of our control. I will come back to these issues when discussing invisible technology, with nanotechnology as an extreme case (section 4). And make a further, and final point about the politics of being part of a world that is being shaped without much control (section 5).

Embodied Expectations

The materialization of the 'not yet', as Barbara Adam has called it,[8] is literally visible in prototypes like early robots which have basic functionalities but cannot do very much yet, and in demos like "the car of the future", and "the kitchen of the future" – with gleaming surfaces indicating modernity/progress –, which mobilize support for further work to materialize them. Similarly, when the US Army develops the Future Warrior it demonstrates the 'Soldier of the Future' with his 2020 Future Warrior uniform system delivering super powers, to USA Congressmen and their staff.[9]

Thus, the embodiment of expectations also structures further developments: mobilizing support which has demands attached to it, articulating the promise into more specific requirements for the next step. While such socio-technical "shaping" of technology does not imply a linear trajectory of increasing performance along the dimensions of functionality originally envisaged, there is structuring of the co-evolution into certain paths (and branchings) and not others.

There are also repercussions deriving from *"effet d'annonce"*. An intriguing example is the case of Snuppy, the cloned puppy.

Researchers at Seoul National University in South Korea announced (in an article in *Nature*, early August 2005) that they had successfully cloned a dog. The cloned puppy was called Snuppy, and a photograph of the puppy with his father (= somatic cell donor) was featured in the scientific and professional press and

---

8   In her project on sociology of the future, www.cardiff.ac.uk/socsci/futures

9   Shown on "Soldier Modernization Day", Capitol Hill, 23 July 2004. See the pictures on www.defenselink.mil/news/Jul2004 (accessed 29 July 2005).

the general media all over the world. The puppy looked cute, not at all a 'hopeful monstrosity' – but if one thinks of it as a technological option, that would be the right term. Snuppy was the only one of 1,095 cloned embryos implanted in 123 dogs to survive to healthy puppyhood.

Now move to the USA, the website of a firm: "**Genetics Savings & Clone** enriches the lives of pet lovers through superior cloning technologies. Cat cloning available today; dog cloning available in 2005."[10] Of course, you have to be assertive on your website. But when the South Koreans announced they had been successful, this implied that Genetics Savings & Clone had to react to the implicit scenario (dog cloning will now be possible).

The Scientist (Aug. 3, 2005) asked GSC for comments: "Phil Damiani, chief scientific officer of Genetic Savings & Clone, which announced in December 2004 that it made the world's first sale of a cloned cat, said the efficiency was "probably one of the lower ones ever done for cloned animals." The company had hoped to be the first to produce a cloned dog and a few years ago had a clone that nearly came to term, Damiani told *The Scientist*. The fetus was alive on ultrasound, but stopped breathing by the time it was delivered by cesarean section. The Korean team has "jumped ahead," he said.

Damiani said that his company remained convinced that their technology–which relies on chromatin transfer, rather than nuclear transfer, and egg and embryo assessment prior to cloning and transfer–would eventually make it possible to clone dogs commercially.

The company expects to be able to produce a cloned dog in the next few months, said Genetic Savings & Clone spokesperson Ben Carlson. "People have been asking us, does this mean that tomorrow you'll be able to start offering this service commercially? We wouldn't be able to make a successful business out of using the technique the South Koreans used," Carlson said. The low efficiency rate, combined with stricter animal welfare rules in the United States that limit the number of times eggs can be harvested and that transfers can be made, would make it impossible. But "it certainly validates our contention that dogs can be cloned," he said. "It doesn't mean we're quite there yet."

Genetics Savings & Clone is forced into explanations, and subsequently extra activities, because of Snuppy. This then determines further trajectories of development, and in the end, the regular existence or non-existence of cloned dogs on our planet.[11] And this is not the final word. The technology and niche demands (of the "pet lovers") continue to co-evolve, and other applications might branch out.

---

10 http://www.savingsandclone.com/index2.html, accessed 5 August 2005.

11 In the recent upheaval about fraudulent human cloning work of Professor Hwang Woo-Suk of Seoul National University, the investigation (by Seoul National University) showing

Co-evolution is also visible in Bruno Latour's well-known analysis of hotel keys (Latour 1991). The hotel manager has to provide the guests with a key to their rooms, but wants the key back in due course. Since guests might forget to hand in their key, he puts a sign on the hotel desk: "Please hand in your key when you leave". Guests are insufficiently disciplined by the sign, so the hotel manager first puts a label on the key with the address of the hotel, so that the key can be sent back, and then makes the key heavier and unwieldy, a material incentive to get rid of the key by handing it in.

Latour is interested in what he calls the 'program' of the hotel manager and the 'counter-program' of the hotel guests who can't be bothered with returning keys. The hotel manager's wish becomes "clothed", i.e. embodied in a heavy key, so gains reality. (ibidem, p. 108) More interesting for our purposes is that the shape of the key is the temporary outcome of the struggle of program and counter-program. The key can stabilize the contestation, but remains open to further changes induced by the programs.[12] So it is part of the ontology of the hotel world, but keeps a prospective component because of its dependence on co-evolving programs.

This example works at two levels: the level of sociotechnical affordances, and the level of paths or trajectories of further development. The hotel key with a weight attached to it offers an affordance for it to be used in certain ways, and not in others. This is how Norman (1990) has discussed affordance designed into artefacts, e.g. doorknobs inviting the user to open the door in the right way. There is also an affordance to further develop the technology in certain directions, and not in others. For a time, increasing the weight and bulkiness of the key appeared to be the only possibility. Thus, the hotel world would be "peopled" by ever heavier keys. By now, electronic key cards offer an alternative, if the hotel manager is prepared to invest in the new locks and in other maintenance.

This links up to the second level of technological paths or trajectories. Phenomena at this level have been analysed extensively by evolutionary economists and sociologists studying technological change. In particular, Nelson and Winter (1977) introduced the notion of what is now called a technological trajectory based on an emerging regime, by pointing out the strength of expectations of engineers and other technology actors about productive directions to go. "… the advent of the DC3 aircraft in the 1930's defined a particular technological regime; metal skin,

---

up the fabrications also concluded that the cloned dog was genuine (Oransky 2006).

12 Latour 1991, p. 110: "the only interesting reality is the shape of the front line". The front line between program and counter-program is also influenced by independent technological developments: key cards, first with perforations coding for the door of the room, then with a magnetic strip, relieved the burden on the hotel manager ("we can easily make a new key card") and allowed further security (change the code of the door by pressing a few keys on the hotel computer).

low wing, piston powered planes. Engineers had some strong notions regarding the potential of this regime. For more than two decades innovation in aircraft design essentially involved better exploitation of this potential; improving the engines, enlarging the planes, making them more efficient." Thus, the airplane world was filled by ever further generations of the DC3 airplane.

Prototypes, and evolving artefacts in general, indicate what they might become. This depends just as much on how actors see, and "read" them, than on what was put into them by their immediate developers. Such "readings" are easy for mechanical contraptions like machines, motor-cars, early and present robots, but more difficult for components and materials that are invisible inputs into visible products. However, such inputs can be made "visible" in diagrams and drawings. One interesting example is a drawing with stem cells in a Petri dish in the middle and (presumably causal) arrows pointing to eventual functional applications. This diagram was featured on the USA National Institutes of Health website for a few years, and was widely reproduced, sometimes without any explanation. It had become an icon. And it shaped discussions about the potential of stem cells, for proponents as well as opponents (of the use of embryonic stem cells).[13]

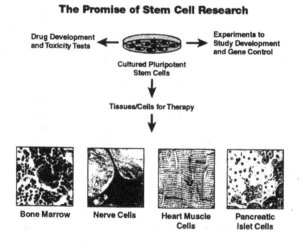

The Promise of Stem Cell Research

Drug Development and Toxicity Tests ← Cultured Pluripotent Stem Cells → Experiments to Study Development and Gene Control

Tissues/Cells for Therapy

Bone Marrow   Nerve Cells   Heart Muscle Cells   Pancreatic Islet Cells

---

13 Indicative is US Senator Frist's speech from the floor of Congress, 29 July 2005, promising support for the pending Stem Cell Research Enhancement Act. A pro-life protagonist from the beginning, now he actually followed the lines of promised applications of the diagram to outline why he was modifying his position.

The possibility of "reading" embodied expectations in artefacts (up to prototypes and demos) and in/through their descriptions and diagrams implies that artefacts can be seen as generalized texts. Grint and Woolgar (1997) have pushed this view, but primarily to show how such "texts" configure the users.[14] I want to pursue the ontological aspects of this view, and consider artefacts as material narratives. The notion of "material narrative" is broader than just an artefact and its embedded script. It also applies to pictures (which tell a story) and landscapes (which guide their "reader" through them).

Material Narratives

Artefacts, in use or on offer, are material "stories" with routing devices (like script, cf. Akrich 1992) and affordances (Norman 1990, Hutchby 2001)) which guide the "reader" without fully determining his/her movements. Think also of "reading" a HiFi installation so that you can work with it – a competence that younger generations have acquired almost without thinking about it. Thus, "reading" of a technology is a practice, building on the affordances, rather than being dependent on conventional texts.[15]

Such "readings" are not limited to the micro-interactions between artefact and user, and behind them, the designer as the "writer" of the original story. Interactions of humans with artefacts, or better, with configurations that work – if interacted with adequately –, take place in an overall sociotechnical landscape and a mosaic of sociotechnical regimes.[16]

---

14 When Woolgar introduced the idea of technology-as-text (Woolgar, 1991), he discussed various ways of understanding "reading" technology-as-text. In the interpretive (or social construction of technology) approach, the variety of possible readings is emphasized. Woolgar suggests this approach implies objective and transcendental existence of a technology, just as in labeling theory in the sociology of deviance it is a matter of different labels for a certain behaviour. His own reflexive approach is ontologically agnostic, which allows questions about how the reality of technology is created, described and sustained (ibidem, pp. 41-42). This "muddies the analytic waters", but that is exactly his purpose. But he then focuses on what this implies for the status of texts of the analysts. Similarly, Grint and Woolgar (1997) emphasize: "What happens to the structure of our [analyst's] discourse when we introduce the notion of machine as text?" (p. 70).

15 Compare also notions of material culture in anthropology, e.g. Appadurai (1986), and for a contemporary slant Miller (1998).

16 The notion of socio-technical landscape overlaps with what Appadurai (1990) called "technoscapes" and Barry (2001) "technological zones of circulation". The key point

Material narratives create agendas, and storylines that continue into the future. Also materially, for example when reverse salients are encountered in the expansion of (large) technical systems that can then be turned into critical problems to be addressed in further work (Hughes 1984). Hughes studied electrical power generation and networks at the turn of the 20th century. By now, there are also electricity grids, their management, and the management of (possible) failures. Our landscapes have become filled up with cables, pipelines for oil and gas, just as well as roads, and now also masts for mobile telephony, related to the "cells" covering the territory. They were shaped by what we wanted to realize, now they shape what we can do, and how we can develop further. [17]

Just as gradients in a landscape (say, hills and valleys) shape the movements of people and other mobiles which traverse the landscape, a socio-technical landscape shapes action and perception. It can be seen as a material story, again with routing devices to guide the "reader" without determining its movements. Some of these routing devices have evolved naturally, and almost all of them are outcomes, at a collective level, of a variety of actor strategies, designs and interventions, and thus to some extent (and after some time) unintended by any of them. It is a *dispositif,* just as much as the more explicitly socio-technical *dispositifs* studied by Foucault and others.[18]

Artefacts are part of such a sociotechnical landscape, and offer avenues of access, or be like boulders that have to be circumvented. The material story is told by the overall set-up, and does not need words of human actors to unfold.

A well-known example of a sociotechnical landscape (but one which needs to be thought through further than is commonly done) are the low overpasses on Long Island, intended by their designer, New York city architect Robert Moses, to keep

---

is that "landscape" is not just a passive backdrop against which humans play out their affairs. It is itself constructed, and part of the "play" is to construct parts of the backdrop.

17 This is very clear in an interesting study of the material unification of the Netherlands (Van der Vleuten, 2003).

18 Cf. the notion of a 'sociotechnical *dispositif*' as introduced by Cohen, Walsh and Richards (2002: 66, 70, 74) after Boltanski and Thevenot. Also Latour (1999: 192): "Purposeful action and intentionality may not be the properties of objects, but they are not properties of humans either. They are the properties of institutions, of apparatuses, of what Foucault called *dispositifs*." Barry (2001, p. 200) comments that "Foucault's analysis of *dispositifs* or apparatuses is too satic to reveal the dynamic instability of socio-technical arrangements."

New York's black and poor white, who had to use busses, away from the beaches and parks he had created on Long Island (Winner 1980).[19]

When you have to travel on a bus, you "read" these overpasses as a barrier. Now that almost everyone has a car, the reading is different. Unless one becomes reflexive, and looks into the history to deconstruct the "naturalness" of this part of the landscape.

Is this just a matter of interpretive flexibility? No, the reader is herself part of a socio-technical ensemble including the overpasses, her having a car or not, and sociotechnical infrastructures and traditions (like being able or not to materially deconstruct an overpass). This shapes the reading.[20]

But such reading is interactive travelling, and modifies the material narrative. It leaves traces, like travellers wear out paths in a landscape, but sometimes also a sudden crosscut. "Reading" is also "writing".[21] This turn of phrase encapsulates the notion of prospective ontology in more foundational way than was possible in the earlier analysis of embodied expectations. Ontology cannot be other than prospective. There is always further reading/writing.

This resonates with a suggestive thought of Ortega y Gasset: "This invented life – invented as a novel or play is invented – man calls "human life", well-being. [...] Have we heard right? Is human life in its most human dimension a work of fiction? Is man a sort of novelist of himself who conceives the fanciful figure of a personage with its unreal occupations and then, for the sake of converting it into reality, does all the things he does – and becomes an engineer?" (Ortega y Gasset, 1962: 108)[22]

---

19 In science & technology studies, this case has the status of an 'urban legend'. This became clear in the discussion of the complexities of the actual story of the overpasses: Joerges (1999a, b)), Steve Woolgar and Geoff Cooper (1999).

20 Here, I endogenize the observing and experiencing subject, against the dominant Cartesian tradition. This approach can be developed further in terms of narrative: the actor becomes the embedded author of an ongoing narrative, where all the characters can turn into authors/actors modifying or contending the thrust of the narrative. This is an actor-network approach, see Callon, Law and Rip (1986). Latour's analysis of programs and counter-programs can be re-formulated in these terms.

21 As Woolgar (1991, p. 37) notes for artefacts: their impact, but also their *de facto* shape is built during the process of design but reconstructed and deconstructed during usage. Thus (subject to the constraints of the actor-network), it is the readers who write the text of technology.

22 Van Lente (1993), after whom I quote Ortega, in the final paragraph of his PhD thesis, turns the thought around: "The reverse is true as well: the engineers have to be like novelists. They have to write forceful fiction and make it come true as well. Actors developing technology, to paraphrase Ortega, conceive for themselves the fanciful figure of a future technology, and for the sake of converting it into reality, they, as well

Engineers, clearly, are seen as "writers". Just as in writing novels, the writer is not in complete mastery of his characters, the unfolding storyline has its own dynamics. Even when the writing is construed as putting building blocks together. Engineers (and other actors involved in creating new configurations) do not always know what they are writing. The configuration might work, but there could be surprises, directly and over time (as I intimated already in the first section).

Invisible Technology

"Reading" (being routed and shaped) and "writing" (inscribing and shaping) are inseparable and constitute evolving reality, whether through experience and actions of humans as part of socio-technical assemblages, or without them. Given this overall take on a process ontology (a post-modern version of pre-Socratic ideas of the basic stuff of the world?), there still is a variety of readings/writings. Interestingly, the material narrative seems easier to read for mechanical contraptions like a motor car than for the components and materials that are invisible inputs into visible products. This is even more striking for nanotechnology, where "incredible tininess" goes together with possible (but precarious) manipulation (Nordmann 2006). At the nano-level, the "writing" is strongly mediated by instruments, and not predictable because one does not know what is happening at the nano-scale until after the fact.[23] Thus, in a first round, invisibility is a matter of being hidden: the nanotechnology input in a sunscreen, or in a coating, is important for its functionality, but not visible to the user. But invisibility goes deeper, creating uncertainties about what is happening "down there", also for the "writers". This then leads on to foundational issues, for example whether our presumed technological abilities to handle invisible components are categorically different from what is labelled as magic.

To start with the superficial (but interesting) point: If "readers" do not see the technology inside, "writers" can develop strategies to make the technology, or at least its contribution to functionalities, visible. I already discussed the example

---

as others drawn into the fancy, do all the things they do. Whatever technology may be in the present, it is rooted in the future."

23   An interesting example is the writing of the IBM logo with xenon atoms on a nickel surface, by Don Eigler of IBM Almaden. The picture is featured on their website, and widely copied. It appears to capture a key feature of nanotechnology, at least is read that way. Similar writings are undertaken, but they continue to be difficult: time consuming because depending on favourable circumstances out of control, like the tip of the probe becoming more pointed because some atoms have fallen off.

of stem cells where diagrams and texts were used to show (and embody) causal linkages between stem cells and eventual functional applications. One simple "writer's" strategy is to announce that there is X (here, nano) inside an artefact. This is how way micro-processor firm Intel created positive visibility by having a label "Intel Inside" put on laptops. However, in the case of engineered tissues, a label "Stem cells were the source" might backfire: it could incite controversy because stem cells derived from embryos continue to be controversial. In the early 1990s, labeling of GM (genetically modified) food seemed a compromise solution for the impasse, because it would shift the question of acceptability to one of consumer choice. Then, (UK) supermarkets decided against having GM food ("Frankenfood") on their shelves at all, to pre-empt possible protests – with the net effect that the compromise was undermined.

For nanotechnology, a strategy of "Nano Inside" will help people appreciate the role of nanotechnology, but may well backfire if something untoward happens elsewhere under the umbrella term "nanotechnology". Prudent companies have come to realize this, and actually shift their strategy. Cosmetics company l'Oréal used to refer to nano in its advertisements, but now speaks of "innovative molecules".

The challenge of invisibility goes further than creating visibility by labeling, i.e. making the inside visible, on the surface, to the outside. The "writing" of the material narrative is involved as well as the "reading". At the nano-scale, seeing is actually feeling (atomic force microscopy depends on a tiny tip moving up and down as it traverses a surface). Turning what is "felt" into images to be seen and read for what they might tell us, is a complicated challenge. Among themselves, nanoscientists talk of "blobbology", the craft skill of interpreting computer-generated images of surfaces. The importance of such craft skills is not exclusive to nanoscience, but nanoscience highlights the interpretative challenges involved.

The dependence of "readers" on the reporting and imaging techniques and interpretations of engineers/technologists ("writers") is further complicated by the fact that the configuring ("writing") is not fully controlled, but delegated to induced self-organisation, say, of layers of molecules on a surface. Will they actually arrange themselves in the desired way? What sort of text will they write? Because of the dominant rhetorics of control, the open-ended character of such self-organisation is backgrounded. But it may reassert itself, when the unexpected occurs.

This has definite implications at the production side, where checks and quality control are needed but can be done only after the fact, and/or where issues of liability and insurance make producers reluctant to sell such "unreliable" products. This implies that, in the end, we may actually not get many of the products that are now being promised on the basis of nanotechnology. The eventual furniture of our world will depend on the implementation of the promises, and the socio-technical constraints and limitations that are encountered.

For the ontology question, the key point is that materials and components do their work (or are expected to do so) while their agency cannot be traced, or not completely . Nordmann (2006) argues that this is a structural feature, not just a matter of something that happens to be invisible now and can be made visible if we put enough effort into it. That is why he introduces the notion of "noumenal technology", playing with Kant's notion of *noumena* and of things *an sich*. For Nordmann, noumenal technology is not only invisible, but out of control because outside our experience.[24]

Nordmann's analysis links up with a general concern, a concern about control: While technology as prospective ontology has always been with us, we may be entering a new phase, given our increased capabilities to interfere, and to a limited extent, control. This was, and is, an issue in genetic engineering, and will be highlighted further by developments in nanotechnology.

Genetic engineering adds to the furniture of the world by modifying what are thought to be the basic building blocks. Nanotechnology claims to be able, or eventually to be able, to manipulate and thus "build the world atom by atom." The rhetorics of so-called converging technologies (nano, bio, info & cogno) suggest that all the basic building blocks (bits, atoms, neurons, genes) are in for manipulation. Thus, a continuing and pervasive vision of control which is pushed in spite of the various practical limitations that can be identified. While the promises may be empty, activities are set in motion which may lead to new findings and technological options. So there will be changes.

Nanotechnology is a domain (or patchwork of overlapping domains) where prospective ontologies materialize. There are promises about the new possibilities at the nanoscale ("there is plenty of room at the bottom" said Feynman in 1959), but these are deeply ambivalent. For example, if nano-size matters (to create new properties and new effects), it also matters in terms of dangers, as with the health and environmental risks of nano-particles.

More important for my argument is the trial-and-error fiddling with matter, at the level of invisible constituents. To obtain interesting effects, self-organisation is induced – but with impredictable effects. The so-called grey-goo scenarios (where tiny nano-robots multiply and turn everything into goo) continue to draw attention, even if they are mythical. But the general concern is legitimate: technology can

---

24 Nordmann further argues that science is in the world of phenomena, and can analyse in terms of causality. Technology as actual interactions, independent of such understanding, partakes of *noumena*. "Technical interventions engage reality" (p. 70).

"escape" or "run away", cf. phrases used in the recombinant DNA and subsequent GMO debate (and also earlier, about nuclear reactors). As Jean-Pierre Dupuy has pointed out, nanoscientists and technologists embrace such ambivalencies, make them part of their job, if not of their skills. They strike out into the unknown – thus, they are intentional sorcerer's apprentices.

Here, another feature of Nordmann's attempt to capture structurally specific characteristics of technology, at least some technologies, as noumenal technology, is visible. Noumenal technology, in his view, belongs with nature outside us, with the brute and uncanny. And that is why there is concern, even fear of invisible technology – even if invisible technology can also be accepted because it delivers good things (Lianos 2003). "[P]ervasive technical interventions change the things-in-themselves, the world not as we know it but where we rely on it unknowingly." (Nordmann 2006: 67) "We cannot trust a *noumenal* technology." (ibidem: 71). Nordmann's and mine question then is whether such a technology can be brought out of the realm of *noumena* into the realm of *phenomena*, and so become part of human experience? We may not have a choice here (if the noumenal is categorically distinct), but the question can still be raised and addressed. Particularly because there are political issues involved (in the broad sense of "political", cf. next section).

One traditional way to capture the noumenal world is through magic. Interestingly, the handling and presentation of the invisible world of nanotechnology may be similar in structure to the practices of magic. Taken in a broad sense again: practices of magic are not limited to shamanic dancing. Prescribing placebos, with demonstrable effects, is benevolent magic just as well.

A leading nanotechnology research centre, IBM Almaden (California), opens its website with the slogan "Working towards the magic of tomorrow's technology". At the bottom, there is a quote from science-fiction writer Arthur C. Clarke (his "third law"): "Any sufficiently advanced technology is indistinguishable from magic." This may be seen as playful presentation, acceptable as part of West Coast culture. But it is serious play. Compare how in the very serious German culture, one of the German nanotechnology research centres opens its website with the same quote from Clarke. Thus, there must be some resonance in the community of nano-scientists & technologists to the notion of magic, as handling the invisible, somehow. One root may be that the handling of molecules at the nanoscale is indistinguishable from the indirect visualisations of the nanoscale: it is always inference about what is happening down there at the bottom.[25]

---

25 This is very visible (if I may say so) in the landmark paper of Hla et al. (2000) depicting induced reaction steps in graphs, images, and visualisations.

Does this imply that the philosophy of technology can (and should) be broadened to include philosophy of magic? One should be serious about magic, anyway, if (and when) it is a practice that performs, somewhat independent from the convictions that accompany it. When the Cartesian idea of control from the outside is left, a range of possibilities of modulating ongoing processes appears (of which Cartesian control is one, and an extreme, version).

Illustrative is the case of placebos that work. They confound so-called evidence-based conclusions from clinical trials, therefore attempts are made to exclude placebo effects, effectively black-boxing them. By now, there is speculation about causalities, e.g. whether placebos stimulate regenerative processes in the body, directly or through the responses of the patient. There is still too little attention to another methodological problem of clinical trials, viz. that it may well be that the drugs being tested disturb regenerative capacities. Anyway, general practitioners are willing to prescribe what they know are placebos, because of the positive effects that appear.

Invisible technologies highlight a general point. Technical interventions engage reality. Their effects are co-produced, and thus difficult to trace back to the intervention as the single source. In other words, such interventions are hopeful rather than determining. There is no categorical difference between Don Eigler struggling to create the IBM logo with xenon atoms on a nickel surface (the only evidence of its being there stemming from the instrument used to create that invisible configuration) and magic practices using other, less technically sophisticated instruments. In other words, Nordmann (2006)'s points about noumenal technology being out of control misses out on such practices of interaction, his view is still predicated on the Cartesian view of control. Instead, going with the flow of the noumenal allows some modulation. Cartesian control, with its projected determinism, then appears as an extreme version of modulation, and one where the principle open-ended character of modulating interventions is pushed out of sight.

Political Ontology

I have articulated a monistic approach, showing the importance of overcoming dualisms,[26] up to replacing ideals of Cartesian control from the outside with being embedded in processes and modulating them as one goes along, "writing" and "read-

---

26 Woolgar (2002) welcomed a "bonfire of dualities", but discussed it primarily at the level of discourse.

ing" at the same time. As in other process ontologies, there is a prospective aspect, and I emphasized this by speaking of prospective ontology, taking open-ended, unruly technology as the entrance point. My discussion of nanotechnology then offered concrete examples and further issues.

Prospective ontology includes projections (and ongoing shaping) of future states of the world as "expected", i.e. possible and possibly desirable. Thus, a prospective ontology is necessarily a political ontology.[27]

A first take on political ontology is ontological politics, the struggle what is to count as "real", legitimate elements of the world. An example from nanotechnology is the struggle about the realism of technological options like molecular manufacturing or nano-robots (as exemplified by the contestation about certain phrases in the USA 21st Century Nanotechnology R&D Act of 2003).[28] Formulated more generally: which human and non-human entities – in a broad sense, including configurations that work (Rip & Kemp 1998) and "ethno-epistemic assemblages" (Irwin and Michael 2003) – are eligible, i.e. deserve, to people the world? Ontological politics is more visible in some areas, e.g. technologies for (and as) human enhancement (Miller and Wilsdon 2006), than in others. The general challenge can be expressed as: if technology is (makes) society more durable (Latour 1991), it is important to artic-ulate the kind of durability that emerges, and perhaps modulate such processes.[29]

A second take on political ontology is that the politics of technology (Winner 1980) where technology is the occasion for traditional politics, have to be replaced by ontological politics (in a broad sense) where technology is the topic. For example, Beck's notion of sub-politics can be extended to the *de facto* sub-politics of artefacts, as these settle debates, implicitly or explicitly. Of course, it is not the artefact *per se* which settles a debate. Artefacts are knots & seams in evolving and irreversibilising

27 The term is used in political science to indicate that it makes a difference which entities one considers to be the basic building blocks of the political world (Hay, forthcoming). Such a use of the term 'ontology' is similar to the use in information science (note 1).

28 An amendment unanimously adopted, called for "a National Academy of Sciences study on the possible regulation of self-replicating machines, the release of such machines in natural environments, the distribution of molecular manufacturing development, the development of defensive technologies, and the use of nanotechnology to extend the capabilities of the human brain" (*21st Century Nanotechnology R&D Act of 2003*, 108th US Congress, First Session, S.189, Section 5 b and c.) There was concern in the nano-establishment: does this indicate that Congress want the Drexler approach to be taken seriously? There was a flurry of letters and telephone calls, including imputations of political wheeling and dealing. Then it died.

29 Later, Latour continues this shift away from "the interminable quarrel over the foundations of the universe" to questions how humans and non-humans "compose the raw material of the collective" (Latour 2004: 61).

so-called seamless webs (Hughes 1986, Rip and Kemp 1998). It is the overall process that counts, in which artefacts can be more or less visible.

In a third take on political ontology, the issue of control returns. If we cannot control for our interference, in general because of unruliness and open-endedness, and for nanotechnology, in spite of its projections of control, because it is based on impredictable manipulations of invisible building blocks, lots of things still happen, but without clear accountabilities. The post-phenomenological analysis of Verbeek's (2005, 2006) importantly adds to this by highlighting the role of human-technology associations, which include "technological intentionalities". In more traditional terms: Who is accountable if control is delegated to self-organized matter? Nanotechnology is perhaps an extreme case, but technology was always unruly and unfinished.

The ideology of control, and as part of technical rationality, has been accepted by many commentators, and this has created intellectual and sometimes political debate about technology which is actually beside the point.[30] Or at least backgrounds and eclipses a much more important point which I have tried to bring out by drawing attention to prospective ontology and unruly technology. "Technological intentionalities" must be part of the conversation, and ongoing repair work is more important than a master plan (or a master configuration that works), which should cater for all eventualities.[31]

I have come a long way since I started looking at the "furniture of the world". I was addressing engineers then, the mandated "writers" of material narratives. My analysis has broadened this picture. Everybody, and everything, is involved. Political ontology is integral to prospective ontology. Prospective ontology opens up the traditional idea of ontology as what is somehow given. So if not given, it might be influenced, modulated. My notion of material narratives, and ongoing "reading" plus "writing", indicates that modulation occurs all the time. Highlighting the aspect of political ontology implies that the *de facto* politics become visible, and more reflexive approaches become possible.

---

30 In a way, technology is victim of the rationalizing tendencies of modernity.

31 Interestingly, recent discussions about governance have undermined the idea of rational control. If it ever worked, it was because of repair work at lower levels in the system. This then implies that some suggestions for productive governance can also be applicable to technology, like the focus on 'spaces' where things can happen, and actors can negotiate. From my analysis of prospective ontology, 'times' must be added: governance includes repair of the future.

Bibliography

Akrich, M. (1992). The de-scription of technical objects. In W. E. Bijker & J. Law (Eds.), *Shaping technology / building society: Studies in sociotechnical change* (pp. 205-224). Cambridge, Massachusetts and London, England: The MIT Press.

Akrich, M. (1995). User representations: Practices, methods and sociology. In A. Rip, T. J. Misa & J. W. Schot (Eds.), *Managing technology in society. The approach of constructive technology assessment* (pp. 167-184). London, New York, NY: Pinter Publishers St. Martin's Press.

Appadurai, A. (Ed.). (1986). *The social life of things. Commodities in cultural perspective.* Cambridge: Cambridge University Press.

Appadurai, A. (1990). Disjuncture and difference in the global cultural economy. *Theory, Culture and Society, 7*(2-3), 295-310.

Baird, D., Nordmann, A., Schummer, J. (Eds.). (2004). *Discovering the nanoscale.* Amsterdam: IOS Press.

Barry, A. (2001). *Political machines. Governing a technological society.* London and New York, NY: The Athlone Press.

Bensaude-Vincent, B. (2006). Two cultures of nanotechnology? In J. Schummer & D. Baird (Eds.), *Nanotechnology Challenges* (pp. 7–28). WORLD SCIENTIFIC. https://doi.org/10.1142/9789812773975_0002

Bergson, H. (1911/1983). *Creative evolution* (translation Arthur Mitchell). Lanham, MD: University Press of America.

Bijker, W. E., & Law, J. (Eds.). (1992). *Shaping technology / building society. Studies in sociotechnical change.* Cambridge, MA: The MIT Press.

Callon, M., Law, J., & Rip, A. (1986). *Mapping the dynamics of science and technology.* Basingstoke and London: Macmillan.

Callon, M. (1999). Whose imposture? Physicists at war with the third person. *Social Studies of Science, 29,* 261-286.

De Laat, B., (1996). *Scripts for the future. Technology foresight, strategic evaluation and socio-technical networks: The confrontation of script-based scenarios.* Amsterdam: University of Amsterdam.

De Laat, B. (2000). Future scripts. In N. Brown, B. Rappert & A. Webster (Eds.), *Contested futures. A sociology of prospective techno-science.* Aldershot etc: Ashgate.

Dupuy, J.-P., & Grinbaum, A. (2006). Living with uncertainty: Towar the ongoing normative assessment of nanotechnology. In J. Schummer & D. Baird, *Nanotechnology challenges* (pp. 287–314). WORLD SCIENTIFIC. https://doi.org/10.1142/9789812773975_0014

Feynman, R. (1960). There's plenty of room at the bottom. *Engineering and Science, 23,* 22-36.

Grint, K., & Woolgar, S. (1997). *The machine at work. Technology, work and organization.* Cambridge: Polity Press.

Hay, C. (forthcoming). Political ontology. In R. E. Goodin & C. Tilly (Eds.), *The Oxford Handbook of contextual political analysis.*

Heidegger, M. (1977). *The question concerning technology, and other essays.* New York: Harper & Row.

Hla, S.-H., Bartels, L., Meyer, G., & Rieder, K.-H. (2000). Inducing all steps of a chemical reaction with the scanning tunneling microscope tip: Towards single molecule engineering. *Physical Review Letters, 85*(13), 2777-2780.

Hughes, T. P. (1986), 'The seamless web: technology, science, etcetera, etcetera', *Social Studies of Science* 16, 281-292.

Hutchby, I. (2001). Technologies, texts and affordances. *Sociology, 35*, 441-456.

Irwin, A., & Michael, M. (2003). *Science, social theory and public knowledge.* Maidenhead and Philadelphia, PA: Open University Press.

IWGN (Interagency Working Group on Nanoscience, Engineering and Technology). (1999). *Nanotechnology – shaping the world atom by atom.* Washington, DC: National Science and Technology Council.

Joerges, B. (1999a). Do politics have artefacts? *Social Studies of Science, 29*(3), 411-431.

Joerges, B. (1999b). Scams cannot be busted. Reply to Woolgar and Cooper. *Social Studies of Science, 29*(3), 450-457.

Latour, B. (1991). Technology is society made durable. In J. Law (Ed.), *A sociology of monsters: Essays on power, technology and domination* (pp. 103-131). London and New York, NY: Routledge.

Latour, B. (1999). *Pandora's hope. Essays on the reality of science studies.* Cambridge, MA: Harvard University Press.

Latour, B. (2004). *Politics of nature. How to bring the sciences into democracy.* Cambridge, MA: Harvard University Press.

Lianos, M. (2003). Social control after Foucault. *Surveillance and Society, 1*(3), 412-430.

Lombardi, O., & Labarca, M. (2005). The ontological autonomy of the chemical world. *Foundations of Chemistry, 7,* 125-148.

Miller, D. (Ed.). (1998). *Material cultures. Why some things matter.* Chicago, IL: University of Chicago Press.

Miller, P., & Wilsdon, J. (Eds.). (2006). *Better humans? The politics of human enhancement and life extension.* London: DEMOS.

Mokyr, J. (1990). *The lever of riches.* New York, NY: Oxford University Press.

Mol, A. (1999). Ontological politics: A word and some questions. In J. Law & J. Hassard (Eds.), *Actor-network theory and after* (pp. 74-89). Oxford: Blackwell's.

Nelson, R. R., & Winter, S. G. (1977). In search of a useful theory of innovation. *Research Policy, 6,* 47-76.

Nersessian, N. J. (2006). The cognitive-cultural systems of the research laboratory. *Organization Studies, 27*(1), 125-145.

Nordmann, A. (2006). Noumenal technology: Reflections on the incredible tininess of nano. In J. Schummer and D. Baird (Eds.), *Nanotechnology Challenges* (pp. 49-72).

Norman, D. A. (1990). *The design of everyday things.* New York, NY: Doubleday.

Oransky, I. (2006). All hwang human cloning work fraudulent. *The Scientist,* published 10 January 2006. Retrieved from www.the-scientist.com

Gasset, J. O. Y. (1962). Man the technician. In idem (Ed.), *History as a system* (pp. 87-164). New York, NY: Norton. Originally published in 1940.

Rip, A., & Kemp, R. (1998). Technological change. In S. Rayner & E. L. Malone (Eds.), *Human choice and climate change* (2nd ed., Chapter 6), (pp. 327-399). Columbus, Ohio: Battelle Press.

Rip, A. (2000). There's no turn like the empirical turn. In P. Kroes & A. Meijers (Eds.), *The empirical turn in the philosophy of technology* (pp. 3-17). Amsterdam etc.: JAI, an imprint of Elsevier Science.

Schon, D. (1983). *The reflective practitioner. How professionals think in action.* New York, NY: Basic Books.

Schummer, J., & Baird, D. (Eds.). (2006). *Nanotechnology challenges. Implications for philosophy, ethics and society.* Singapore: World Scientific Publishing Co.

Stein, R. L. (2004). Towards a process philosophy of chemistry. *Hyle – International Journal for Philosophy of Chemistry, 10*(1), 1-17.

Stoelhorst, J.-W. (1997). *In search of a dynamic theory of the firm. An evolutionary perspective on competition under conditions of technological change, with an application to the semi-conductor industry.* Enschede: University of Twente.

Tsoukas, H. (2005). *Complex knowledge: Studies in organizational epistemology.* Oxford: Oxford University Press.

Van den Belt, H., & Rip, A. (1987). The Nelson-Winter/Dosi model and synthetic dye chemistry. In W. E. Bijker, T. P. Hughes & T. J. Pinch (Eds.), *The social construction of technological systems. New directions in the sociology and history of technology* (pp. 135-158). Cambridge, MA: MIT Press.

Van der Vleuten, E. B. A. (2003). De materiele eenwording van Nederland. In J. W. Schot, H. W. Lintsen, A. Rip & A. A. A. de la Bruheze (Eds.), *Techniek in Nederland in de Twintigste Eeuw. VII. Techniek en Modernisering. Balans van de Twintigste Eeuw* (pp. 43-73). Zuthphen: Walburg Pers.

Van de Ven, A. H., & Poole, M. S. (2005). Alternative approaches for studying organizational change. *Organization Studies, 26*(5), 1377-1404.

Van Lente, H. (1993). *Promising technology – The dynamics of expectations in technological developments.* Enschede: University of Twente.

Van Lente, H., & Rip, A. (1998). Expectations in technological developments: An example of prospective structures to be filled in by agency. In C. Disco & B. J. R. van der Meulen (Eds.), *Getting new technologies together* (pp. 195-220). Berlin: Walter de Gruyter.

Verbeek, P.-P. (2005). *What things do. Philosophcal reflections on technology, agency, and design.* University Park, PA: Pennsylvania State University Press.

Verbeek, P.-P. (2006). Materializing morality—design ethics and technological mediation. *Science, Technology & Human Values, 31*(3), 361-380.

Walsh, V., Cohen, C., & Richards, A. (2002). The incorporation of user needs in telecom product design. In A. McMeekin, K. Green, M. Tomlinson & V. Walsh (Eds.), *Innovation by demand. An interdisciplinary approach to the study of demand and its role in innovation* (pp. 168-186). Manchester and New York, NY: Manchester University Press.

Whitehead, A. N. (1929). *Process and reality.* London: Macmillan.

Winner, L. (1980). Do artifacts have politics? Reprinted In D. MacKenzie & J. Wajcman (Eds.), *The social shaping of technology* (2nd ed.). (pp. 28-40). Buckingham & Philadelphia, PA: Open University Press.

Wood, M. (2002). Mind the gap? A processual reconsideration of organizational knowledge. *Organization, 9*(1), 151-171.

Woolgar, S. (1991). The turn to technology in social studies of science. *Science, Technology & Human Values, 16*(1), 20-50.

Woolgar, S., & Cooper, G. (1999). Do artefacts have ambivalence? Moses' bridges, winner's bridges and other urban legends in S&TS. *Social Studies of Science, 29*(3), 433-449.

Woolgar, S. (2002). After word? – On some dynamics of duality interrogation. Or: Why Bonfires are not enough. *Theory, Culture & Society, 19*(5-6), 261-270.

Wynne, B. (1988). Unruly technology: Practical rules, impractical discourses and public understanding. *Social Studies of Science, 18*, 147-167.

Chapter 8
Interlocking Socio-Technical Worlds*

As part of the procedure when I applied for the chair of Philosophy of Science and Technology at the University of Twente, I had to give a presentation about a topic of my own choice. So on April 13, 1987, the Appointment Committee could listen to me presenting 'Repertoires en gekoppelde sociale werelden'. In my introduction, I noted that 'construction of technology' could be the central theme for research in De Boerderij group, and for teaching in the WWTS degree course (WWTS = Philosophy of Science, Technology and Society). There would, of course, be a link to Social Construction of Technology (the SCOT approach), but that would be only one element. Constructive TA would be another, and interaction with technological courses a third. The theme would thus include present questions and approaches, but also lead to further theory development. It is the latter that I then wanted to focus on in my presentation.

When organizer Willem Halffman suggested I would use the STeHPS collo-quium in June 2006 to look back to the first lecture (on Science, Technology and Society) I had given, I agreed but it seemed more interesting to go back to the first lecture I had given in Twente. I remembered having kept the written-out notes of a lecture I thought I gave in October 1987. But it turned out to be this lecture for the Appointment Committee. Re-reading my notes, I was surprised how much of my present thinking was already visible in it. An indication of how I have been reiterating myself all these years? There is some progress, though. At least, I now speak of socio-technical rather than social worlds.

In the following, I present a version of that lecture, keeping to the structure of my presentation in 1987 but adding further material and analysis, and some further thoughts (in footnotes, and also in the main text). The original text (i.e.

---

*   June 2006 (slightly revised, May 2011)

© Springer Fachmedien Wiesbaden GmbH, part of Springer Nature 2018   157
A. Rip, *Futures of Science and Technology in Society*, Technikzukünfte,
Wissenschaft und Gesellschaft / Futures of Technology, Science and
Society, https://doi.org/10.1007/978-3-658-21754-9_9

the revised English version) is recognizable as highlighted in italics, to satisfy the history buffs (including myself). The Figures are power point versions of the original transparencies, except for the one with Sahal's landscape.

The SCOT approach (Bijker and Pinch, but actually the general approach of the Bo-erderij group linked to Social Integration Theory) is an important step in the analysis, and hopefully also the explanation, of technological developments. But it is not the whole story, for example because long-term and structural aspects cannot easily be addressed. But we can identify the next steps.

As entrance point – and a crucial aspect of present-day technology – I use the phenomenon of strategic interaction with and around technology, for example in technology policy, in firm strategies, but also visible in the recognition of the im-portance of a technology assessment "philosophy". This implies that a new factor is becoming important in technological developments: strategic technology policy [and management]. This entrance point is additionally important because analyses taking it into account have immediate implications for change strategies and policy advice.

Strategic technology policy is not a purely external factor: technology actors actively take possible policies into account, try to influence or even initiate particular policies, etc. In other words, there are <u>coupled circuits</u>. Figure 1 visualizes the interactions.

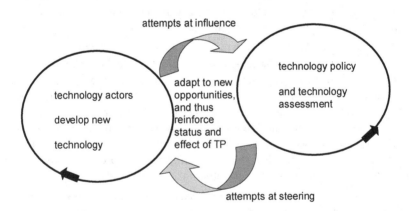

This sketch is not just a visualization of a particular set of interactions. It is part of a more general theory, in any case a theory of development of science and technology in context, in which micro-processes can be coupled with meso- and macro-processes

and structures. For patronage of science, for example through government-related funding agencies, I have articulated such sketches in detail and applied them to analyse structures and processes.[1] In other words, there is at least one case for which it is a viable theory.

The theory should be formulated and visualized so as to allow linkages between circuits in general, not just between policy and a domain that is addressed by the policy. *Key elements of the theory are dynamic stabilization through (1) emerging repertoires, and (2) building and maintenance of 'social worlds' – Anselm Strauss's concept, not too different from the concept of 'game' in social integration theory.*

In passing I note that these key elements are also a productive starting point for reflection and discussion about issues (including goals and their realization) of technology policy and technology assessment, grounding them so as to overcome an exclusive focus on principled discussions which remain symbolic because conducted without reference to contexts and dynamics.

In 1987, in the introduction to my presentation, I further characterized my presentation as work-in-progress, and at the same time ambitious. That might then lead to interesting discussion with the committee. My choice of topic was motivated by the fact that I had already presented some elements (see Step 4) at a conference in Brighton in 1985, where Peter Boskma also gave a plenary lecture. He and I concluded our approaches were very similar, and planned to discuss them further. That never happened. Presenting my approach now, in 1987, could thus be seen as a *hommage* to Peter.

Step 1

I will start discussing repertoires.

By now, it is easy to say, in debates about risks of nuclear energy: "After Chernobyl, we must ...". We can short-circuit a complex argumentation by referring to "Chernobyl". Interestingly, while the accident at Chernobyl is located in time (30 April 1986), we do not refer to a date, but to a place. We can use "after" because we refer to a demarcation with the past (as Prime Minister Den Uyl famously did

---

1   references.

in a speech in Nijmegen during the oil crisis in 1973: "That time (world) will never return"). And we use "Chernobyl" as a label to stand for a complex of experiences and arguments which are now blackboxed. It has become an 'ideograph' (McGee), or a point representation (Callon, Law and Rip) and this can be indicated by using capital letters: CHERNOBYL. *What exactly happened, and what should be the message, is not articulated any more. The fact that it is a point representation of a black box is what exerts force.*

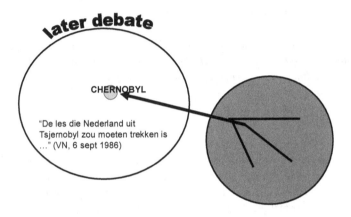

The structure visualized in Figure 2 is identical to that of a spokesperson <u>representing</u> *an institution or a group – a social world. Blackboxing and point representation allows for* <u>circulation</u>. *Ideographs are available without much effort in a repertoire while they can still be used for different purposes. Spokespersons allow social worlds to be present elsewhere, even if they have to translate (and in that sense misrepresent) their social world into a point representation in another world.*[2]

---

2 *Here, I am equating keywords in a text with spokespersons for a world. In the social-semiotic version of actor-network theory both are the outcome of translations. The equation is not simple, though. Consider for example situations where a would-be spokesperson (say, for the environment) uses ideographs (like SUSTAINABILITY) to create a position for himself/herself. If one looks in detail into the processes leading to blackboxing and point representation, one sees projected actor-worlds (embedded in the content of generalized texts) in which roles are assigned for the various characters. Each character can become the author of a competing actor-world, positioning the author of the earlier actor-world as*

The force of a point representation resides, in the first instance, in the represented world remaining blackboxed. Compare the insecurity that results from deconstruction (or just history) of scientific facts, showing their interpretive flexibility: the facts seem to lose their force.

Repertoires contain point representations, from ideographs to stereotypes, standard roles, and attributions and rules. Their specific force derives from these point representations and how they are mobilized and used. Scientific repertoires contain scientific facts, paradigmatic approaches and evaluation criteria. The latter overlap with criteria used in scientific journals and funding agencies, where evaluative repertoires emerge suited to the task of assessing manuscripts and research proposals. In society, there is a patchwork of repertoires, containing generally accepted views, stereotypes, roles, rules and demarcations. Following such repertoires allows one to be a member of society (an insider) and not having to put a lot of effort into all actions and considerations. In other words, routines are not just behavioural routines of an individual, as psychologists would have it. They are ways to survive in worlds which offer such routines.

Step 2

The examples I gave of point representations, from CHERNOBYL to scientific facts (FACTS) are all used in <u>more or less public repertoires</u>, and the translations involved in the emergence of such point representations attempt to de-contextualize the local and contingent. They become general and mobile by becoming inflexible. In local, daily practices, on the other hand, there are <u>informal repertoires</u>, not for outside consumption and thus allowing negotiation about the de-contextualization, and keeping visible what went into it. They are still repertoires, full of accepted cultural rules which newcomers must learn: "This is how we do things here." Newcomers are seen as apprentices. Cf. also David Boje on 'terse stories' in an organization.[3]

---

a character in his. When these translation add up to a more or less shared and stabilized picture, there is a degree of blackboxing.

3   [From my book manuscript Ch. 3] cf. Marsh et al. (1978, at p. 19): "Most institutions offer an official life trajectory in terms of which honour can be acquired and reputation gained. (..) Most institutions also have an underlife, defining other forms of moral career with other forms of hazard and other ways of rendering honour and marking reputation." When David Edge (1979) refers to the "weak underbelly of science," he is also trying to show the "total" system, not just its official face: Edge criticized the idea that formal communication in science is continuous with informal communication and local practices: "[if] formal

An interesting example from the world of science is the list of examples of "what he said" (as author of a published article) and "what he meant" (what really happened in the lab and in the office). The Table below has some examples from the list. What is additionally interesting is how such lists are copied and circulate, are put on the notice board in the lab. The contrast between the informal repertoire and the public repertoire becomes itself part of the informal repertoire. For the public repertoire, the contrast is a risk, it might undermine the power of the point representation. It is sometimes suppressed – cf. the story of Van Kampen at Saris' inaugural lecture in Utrecht. Or it is individualized, as when massaging the data, endemic in science but possibly escalating into fraud, is positioned as a deviation from the norm, and from regular practice, caused by moral weakness of the guy and the external pressures on science.

"What he wrote"	"What he meant"
'Three of the samples were chosen for detailed study.'	'The results on the others didn't make sense and were ignored.'
'Thanks are due to Joe Glotz for assistance with the experiments and to John Doe for valuable discussions.'	'Glotz did the work and Doe explained what it meant.'

From Weber and Mendoza (1973)

Gilbert and Mulkay have addressed the same phenomenon in terms of two repertoires. Adapting their terminology slightly, there is a contingent repertoire *in which all the local practices, the contingencies and the (social) construction of scientific facts can be discussed, and a* rational repertoire *where argumentative coherence of "cleaned-up" data are foregrounded. Scientific articles draw on the rational repertoire, except when there is a heated controversy and authors allow themselves comments about insufficient practices and biases of the other party. In other words, while there may be mixes of the two repertoires, they are used on different occasions. Among insiders,*

---

communication in science is 'the tip of the iceberg' (..) there is every indication that the 'tip' is *radically different in kind* from what is 'below the waterline'. (Perhaps 'the soft underbelly of science' might be a more appropriate metaphor!)." (at p. 114). Mars (1982) analyses the "total reward system" in jobs, which includes informal rewards and fiddling, and points out, in the summary of his case studies, that the first six months, in quite a number of occupations, are necessary to adjust to the mechanics of fiddling in these occupations and more important, of the *need* to fiddle and of the self-acceptance of the role of the fiddler (at p. 193).

and sometimes in interviews, scientists draw on the contingent repertoire, but for more public occasions it is the rational repertoire. (Cf. also Altimore who found such a contrast between what molecular biologists said at hearings about recombinant DNA, and what they said in interviews he conducted with them.)

Thus, it is not just a matter of having more than one repertoire available. The two repertoires are linked, because they refer to the same range of phenomena and activities. And there are constraints on their use. In particular, the contingent repertoire is to be limited to insiders. It functions <u>within</u> a social world, contributes to its coherence,[4] but should not be used for, let alone by, outsiders. (Compare jokes, as about Catholic priests – which can only be made by Roman Catholics, and often only among themselves.)

I add two recent examples of such contingent and rational repertoires. The first is from a study of animal ethics practices and views in an immunology lab.[5] The immunologists can always offer the utilitarian ethics argument to justify their use of animals as models: the suffering of these animals is acceptable because it makes a larger good (improved health for many humans) possible. There are additional arguments about minimizing the suffering of the animals, and sometimes also about respecting their integrity. This clearly is part of the rational repertoire (and blackboxes the fact that lots of the research will turn out to be not directly relevant for the relief of human suffering). The laboratory ethnographer also noted how killing the animal (in this case, a mouse) is presented as an act of mercy – but also as a relief to the scientist. "As soon as the animal is dead [i.e. does not struggle and bite anymore] then I can work in peace, then I can concentrate what I do [preparing tissues for analysis]." This is part of the contingent repertoire, and can be shared with the ethnographer. When the ethnographer presented such items in an internal seminar of the lab, the scientists nodded, this is how it goes.[6]

---

4   This is what Wenger's et al. notion of 'communities of practice' emphasizes. There is little attention to the boundaries of such a community (and insider/outsider attributions), that is why I prefer the notion of 'social world': a 'world' has a boundary.

5   Daniel Bischur (University of Salzburg), *Social Frames of Ethics in the Scientific Practices of the Life-Sciences*, paper presented at the meeting of the International Sociological Association, Durban, 24-29 July 2006.

6   A further example from the same study is how the immunologists draw a line, saying they are prepared to work with mice, but "we don't like to work with [monkeys]". They restrict themselves to lower animals, and can offer such a restriction to show they have reflected on the ethics of animal experimentation. But when there is some interesting phenomenon to pursue, they are willing to lift such a restriction.

The second example introduces further complexities, because of the presence of more parties, with their own interests and their own sociotechnical worlds. It is about tests and the occurrence of false positives and false negatives. This is recognized for hypothesis testing (errors of the first and of the second kind), and styles develop in labs, and in research areas to make an effort to avoid errors of the first kind, or conversely, to avoid errors of the second kind. The same difficulties arise for medical tests (and tests of materials and devices), but then there are more parties involved, and taking action on a false positive (or inaction on a false negative) has further consequences. In a study of HIV-Aids tests,[7] it became clear that further interpretation of the results is necessary to avoid acting on false positives. Clinical and behavioural data (is the person a drug user? a gay person?) is introduced to decide whether to accept the positive test, or go for further tests. These are common practices, and algorithms have been developed to guide the decision steps. In that sense, there is a rational repertoire which covers the practices, while the actual interpretive decisions remain in the domain of the contingent repertoire.

Interestingly, there is reluctance among the scientists involved to let the interpretive element in the tests become widely known. They prefer to present the tests as objective and definitive, otherwise the screening programs would be endangered.[8] In other words, the actual practices should be kept to the insiders' world.

For other actors, in particular companies selling such test and risking liability claims (particularly in the USA), the situation is different. Organon Teknika says of its antibody test kit: "Do not use this kit as the sole basis for diagnosis of HIV." And Abbott Labs is even more explicit: "False positive results can be expected with any test kit." By now, also patients and patients' organizations mix in the fray, and pursue the limitations of the tests from their perspectives.

The original examples of contingent and rational repertoires were from the protected space of the lab, and the need to decontextualize in order to get articles published. With medical tests that are (or are intended to be) widely used, there no such easy protected space, but the scientists are involved in the same dynamics and are concerned about their professional status. Companies, on the other hand, are sellers of products on an open market, need to protect themselves, and this implies pointing out limitations on the use of their products.

---

7   Kevin Corbett (Liverpool John Moores University), *Inside the' black box' of the antibody test: deconstructing official classification of 'risk' in the test algorithms used for identifying the human immunodeficiency virus*, paper presented at the meeting of the International Sociological Association, Durban, 24-29 July 2006.

8   This is particularly the case in the UK. When Kevin Corbett published an earlier critical analysis of the use of testkits in *Practising Midwives* (1999), there were angry reactions from scientists, exactly on this point.

I will have to come back to such complexities. For the moment, the important point is, first, that there are contingent and rational repertoires, and second, *that the occurrence of the two linked repertoires is a general phenomenon. For that reason, I will speak of C-repertoires and R-repertoires, and thus be able to draw on the Gilbert & Mulkay terminology, but take away the specific reference to scientific practices. (i.e. I de-contextualize the approach to make it applicable to other domains.)*

Step 3

The next step (not taken by Gilbert and Mulkay) is to consider the linkages between C- and R-repertoires.

In the worlds of science, there is a struggle, conducted in terms of the C-repertoire, to obtain the status of a "fact" that can be part of the R-repertoire. This struggle, and the terms and conditions under which it is conducted, is itself the outcome of historical processes.[9] *Latour (with Woolgar) has analysed such struggles in terms of a "ladder" of modalities of increasing facticity, from 'speculation' and 'preliminary findings' to received facts, where there are movements up (the construction of facticity) as well as down the "ladder" (the deconstruction of facticity).*

Latour's as well as my point is that this struggle is what makes for the productivity and quality of scientific practices. To be effective, however, the nature/content of the R-repertoire must be somewhat independent from the C-repertoire struggle, other-wise it would continually shift with the vicissitudes of the struggle (as it does to some extent: the victors in the struggle will define what the criteria of facticity and scientific quality are). Such independency has two sources. One, because R-repertoires are public and non-local, individual actors cannot change the R-repertoire as they wish. Two, because scientific worlds have external linkages, e.g. through societal acceptance of their mandate to do science, relatively protected from outside interference. Even the collectivity of scientists cannot change elements of the R-repertoire which are part of such a mandate.[10]

---

9   Cf. my book manuscript, Chapters 2 and 3.

10 *[From my book manuscript Ch. 3] Items in the R-repertoire cannot be modified at will, because they link up with the wider world; nowadays in two steps: the cosmopolitan world "inside" science, and the wider world more generally. At the same time, items from the R-repertoire like the need to fully support a knowledge claim, and norms like organised scepticism and disinterestedness, have a function within the C-repertoire. Even if they do not determine action directly, they are part of the frameworks of negotiation among the actors.*

External anchoring of science as less to do with a reality external to us to which science has access, somehow: 'reality' is always mediated, and very strongly so in laboratories where 'reality' is actively created, as a 'lab world' (Hacking). The mandate might well refer to 'natural enquiry', but is shaped by links with society, in particular powerful actors in society. The 'new learning' of the 17th century was able to find a space by claiming it could offer insights independent of power and tradition, and the struggles linked to them. It was this external negotiation which introduced an R-repertoire, and an R-repertoire in which claims of neutrality and objectivity were keystones, because they were linked to societal support.

I note, for later reference, that the question of productivity and quality raises further issues. Without an R-repertoire, the practice could shift, and drift in any direction. With an R-repertoire, there are constraints. While constraints are necessary (to avoid slipping), some constraints can be better than others. What would be a "good" R-repertoire? The subsequent, second normative question is then about productive interaction between a (hopefully) good R-repertoire, and C-repertoires – see also Step 4.

Whatever the precise linkages, there clearly is a <u>dual relation</u> between C- and R-repertoires. In the worlds of science, the dual relationship can be traced in detail, but the relationship occurs in many other situations. And especially when there is some (proto-) professionalisation, because professionalisation implies some de-contextualisation and circulation, as well as the need for a mandate. Three examples are the construction of new technology, consideration of risks of technology, and implementation and enforcement of environment and safety regulation. The third example allows me to introduce further theoretical points, so I will discuss it separately in Step 4.

In constructing new technologies, local factors are dominant in the early stages. Reports and discussions draw on a C-repertoire. Over time, there is stabilization

---

Kemp (1977), for example, has shown how each of the Mertonian norms is invoked in the circular letters exchanged in a controversy among biochemists, and exerts some force, even if never conclusively. In addition, items from the R-repertoire function not only in informal conversation and exchanges. They are part of the working life, helps it to make productive. And it plays a key role in knowledge production. The struggle for facticity is fought with the help of the two coupled repertoires. This helps to solve the riddle how science, so keen on achieving knowledge, can be based on the quicksand of knowledge claims which are literally untrue (because transcending immediate findings to make a claim). Without an R-repertoire, the practice could shift, drift, in any direction. With an R-repertoire, there are constraints. While constraints are necessary (to avoid slipping), some constraints can be better than others. What would be a "good" R-repertoire?

(and reification), variation and interpretive flexibility are reduced. An R-repertoire has emerged, now including blackboxed artifacts which are what they are and can circulate across locations.[11] The C-repertoire does not disappear, but shifts from the contingencies of developing a new technology to the contingencies of re-contextualization on location (cf. the difficulty of getting a newly acquired technology to actually work), and the modifications that turn out to be necessary.

Risks of technology include 'normal abnormalities' (Brian Wynne's term). Working with a technology implies, unavoidably so, the development of routines to handle it in practice, to cut corners, to keep the practice doable. Such informal routines may deviate from the official design, control and maintenance guidelines (part of the R-repertoire), and can thus be labeled 'abnormal'. But they are 'normal' in the practices that have developed, and may well be productive. Until there is an accident ... This is how one can interpret the responses to the Challenger catastrophe. It is easy to blame engineers and/or managers for the catastrophe when they did not follow the official guidelines, and create heroes (after the fact) out of the early warners (in this case, engineers concerned about blackening of the O-rings, whose concerns were not listened to). Part of the C-repertoire, however, was the experience that O-rings had been blackened without causing any problem, so it could be considered to be nothing special (nothingunusual).

Particularly in this second example, the need for dynamic quality control, *and identification of opportunities for improvement becomes visible. My point is that such improvements have to be sought in the content and interaction of C- and R-repertoires, rather than in the usual approach of blaming individuals for failing. I have argued for this already when discussing fraud in science, which is blamed on the individuals – because science cannot fail. For technology, the point is more pressing, because unlike in science, accidents will have real victims.*

Dynamic quality control does occur, but its heroes are often unsung. This is visible in my third example, which will take social worlds into account explicitly.

Step 4

Many industrial firms comply with environmental regulations. It is is not obvious that they will do so. If the profit motive is what drives them (and managers and CEOs tend to refer to the need to make a profit, if only to close difficult discussions about

---

11 Cf. Latour on 'immutable mobiles', and for further analysis of decontextualization, Jasper Deuten's PhD thesis on Cosmopolitanization of Technology.

what the firm should do), why would they ever comply to environmental regulations, e.g. about handling wastes? Violations (infringements) of the rules will not be discovered (in the 1980s, in only 25% of the cases), and when found out, the sanctions are light and/or can be postponed and reduced by protests and court cases. A simple cost-benefit calculation would drive firms to not, or only minimally, comply with regulations. Indeed, there are so-called 'cowboy firms' which do not comply at all, and enjoy the benefits. If they are caught, and face strong action, they might just end their business (and continue their operations elsewhere, under another name). But the majority are 'good firms', who want to avoid scandals, and pride themselves on maintaining good relations with environmental inspectors.[12]

What happens is that the good firms are part of <u>a new social world</u>, together with the environmental inspectors out on duty in the field. It is an enforcement [and compliance] world in which a productive C-repertoire has emerged, with linkages to the R-repertoires of the firms (profit motive) and the inspectorate (enforcement of regulation). The C-repertoire allows inspectors to forget about enforcing the rules and focus on avoiding environmental pollution in practice, in exchange for "good" behaviour of the firms they interact with. "Better a dirty conscience than a dirty world" is their motto.[13] *That it is a social world is clear from the occurrence of inclusion & exclusion moves. Firms which go against the informal rules are labeled as deviants ("cowboy firms"), and are treated harshly. "Good firms" on the other hand can have their occasional waste problem, but it is then treated as an unfortunate accident. At the side of the inspectorate, there are also inclusion/exclusion pressures. They have to avoid strict adherence to the rules, or will be seen as fanatics (also by their colleagues) and disavowed.*[14]

---

12 *'NRC Handelsblad', 29 April 2006, presented commercial stem cell therapy ("trade in hope") under the heading: 'Stemcell cowboys conquer the world, now also in Rotterdam'. On the front page, it referred to "dubious treatments", and used as heading: "Specialists call for a stop of 'stem cell pirates". The "good" therapists were trying to exclude the "cowboys".*

13 *Or <u>was</u> their motto. Changes in governance (partly inspired by new public management) have forced inspectors to become more distantiated, with fewer interactions in the field. A striking example is the Dutch Occupational Health & Safety Inspectorate, which is now limited to advise on health & safety activities in general, rather than interact with firms.*

14 *'Policy Studies Journal', theme issue "Cross-National Comparisons in Environmental Protection: A Symposium," September 1982. In a draft paper for a summer workshop at IIASA, July 1983, I noted: There is a contingent repertoire (C) which talks of bargaining, of being pragmatical, "the 'twilight zone' process of seeking voluntary compliance and negotiating stipulations" (59) [Numbers between parentheses refer to pages of Policy Studies Journal.] There is also a rational repertoire (R) where enforcement is seen as the execution of the rules, "command" instead of "bargaining" (139), the "strict liability" that does not take into account the intentions of the actors (160) or the possibility of "accidents" that decrease culpability.*

Social worlds are characterized by dual (C+R) repertoires, in general, but this also drives newly emerging "bridging" social worlds, like the enforcement & compliance world which bridges firms (certain departments and individuals in firms) and inspectors.[15] In the C-repertoire, negotiation and interpretation are central. The inspector checks whether the firm remains within certain levels of compliance, and interprets violations as accidents for which the firm can be excused. The inspectors can refer to the official regulations and sanctions – that is their R-repertoire – , but only as a last resort. If they would apply the regulations literally, that might lead to system-wide protests and refusal (cf. civil disobedience?), and thus be improductive. Firms have their R-repertoire of profit maximalisation for their shareholders, and guaranteeing continuity of the firm. Again, literal application of this R-repertoire may be counterproductive. Compare how firms like the ITT conglomerate are disavowed as focusing on profit only.

These social worlds, and social worlds in general, depend on the productive duality of their C- and R-repertoires, but also on their external links, through their R-repertoire or otherwise. Environmental and safety staff in firms create a link with other departments in the firm and with overall management.[16] Inspectors out in the field return to their office in the government ministry, and have to justify their actions there.

There are two important points to note about the relation between the two repertoires. (1) Inspectors and other enforcement agents do not see it as their task to enforce the law. Instead, their goal is to protect the waste treatment system from harm, to solve effluent problems (158), to contribute to an adequate solution to the pollution problem faced in a given case while minimizing enforcement costs (139). (2) The enforcement of the law is a resource in the enforcement process, not an end in itself (163).
Health and Safety Inspectors: interviews AR.
Court rooms: Atkinson & Drew 1979.

15 *Can be linked to the notions of 'trading zone' (Galison) and 'transaction spaces' (Nowotny, Rip, Garraway).*

16 *See our analysis of the social world of safety staff in a firm, and scenarios of its further evolution (Rip & Heitink 1988).*

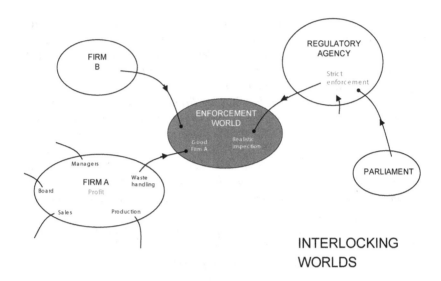

INTERLOCKING
WORLDS

The enforcement world functions between two extremes ('poles'). One extreme occurs when the links to other social worlds are completely backgrounded, and interactions within the world are the focus. The inspectors as well as the staff of the firm "go native", their allegiance is to their shared world. This is also a way to operationalize trust. In the other extreme, the inspectors and the staff of the firm are spokespersons for their respective worlds, and they interact strategically. Since they are bound together (mutual dependency), one can speak of a strategic game and see their actions as moves in a strategic game. This is actually how the actors will interpret actions of others – who are players rather than members.

In practice, shifting mixed or compromise arrangements occur, and oscillate between the two poles. An interesting example is how inspectors construct a gradient of force to keep the firms in line, and can do so only when the firms need to remain members of the shared world, i.e. be good firms.

Graded persuasion in the enforcement process

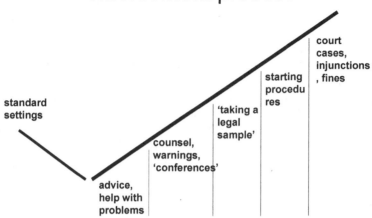

disturbing the system – may lead to protests and refusal (cf. civil disobedience)

[Mention 'stem cell cowboys' again here?]
[This can be developed further in terms of productivity of dual repertoires and their embedment in structures (Step 6). What I would now call governance. But first, more about structure.]

Step 5

Linkages connect social worlds, and these linkages cannot, therefore, be changed easily from within one or another social world. Furthermore, the dual repertoires enable and stabilize the linkages between these worlds. In that sense, there is <u>structure</u> *(at the meso-level), and a structure which can force what happens in the linked social worlds.*

This approach ('theory') is dynamic. Its thrust is similar to Giddens' structuration theory, but pays more attention to meso-level structuring in terms of interlocking social worlds. Norbert Elias' work on insiders and outsiders, and on configurations in general, is clearly relevant as well.

An additional advantage is the possibility of addressing multi-level phenomena, and the dynamic interactions across levels. Instead of stipulating that there are certain levels, somehow, as it were given by nature (i.e. society), we can now see higher levels as the outcome of de-contextualisations leading to a cosmopolitan level (e.g. of a scientific field or a technological regime), and/or induced by government actions delegating tasks to intermediary actors (e.g. funding agencies) which take on a life of their own. In other words, 'levels' do not exist as such, but are products of decontextualisation and recontextualisation in a new social world full of point representations of other social worlds.

The macro-level might be characterized as shared and somewhat forceful references (cf. Karin Knorr in Knorr & Cicourel) rather than an identifiable social world. But our societies are hierarchically constituted, with authoritative actors at the collective level, which can be seen as a macro-level (cf. also Rip 1995 on macro-actors in relation to introduction of new technology). Such authoritative actors, e.g. government, function in their own social worlds. But they derive their authority towards other social worlds from their mandate, especially their link to accepted formal and *de facto* constitution of our society.

Thirdly, socio-technical landscapes shape actions through the gradients of force they introduce, somewhat independent of ongoing decontextualisations and recontextualisations. When generally accepted, they act as a *de facto* sociotechnical constitution.

Macro level?

- For technology, we have developed analysis in terms of regimes (the grammar shaping technological development in a domain), and an evolving patchwork of regimes

- Against the backdrop of sociotechnical landscapes

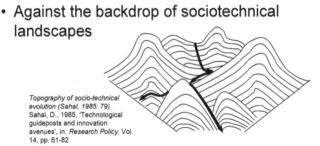

Topography of socio-technical evolution (Sahal, 1985: 79) Sahal, D., 1985, 'Technological guideposts and innovation avenues', in: *Research Policy*, Vol. 14, pp. 61-82

What we have now is a three-level conceptualization. A conceptualization which I have often presented in terms of micro, meso and macro, while feeling guilty about the simplifications (as to increasing scope, and as to hierarchy) that this terminology implies.

- The first level ('micro') has individual actions embedded in practices, more or less stabilized by social worlds.
- The second level ('meso') consists of interlocking social (socio-technical) worlds, where the interlocking adds up to a structure that cannot be influenced easily by actors. The interlockings have horizontal and vertical dimensions, so the meso-level is a complex patchwork.[17]
- The third level ('macro') is more a background to this patchwork than that it sits above it. It is a backdrop – a sociotechnical landscape – which enables and constrains, as in Fernand Braudel's *longue durée*.]

While scientific fields and funding agencies have their own social worlds, with dual repertoires of their own, they enable and constrain what happens on location, in scientific practices.

[Since people move about and cross into other worlds, there will be links, already by attribution. For example, in the C-repertoire of scientists there can be talk about assessment of proposals by funding agencies as being a lottery, and some scientists can report on their experience as members of a (peer) review panel. In the C-repertoire of actors in the world of the funding agency, there is room to recognize elements of lottery. Important elements of the C-repertoire are the notion of 'proposal champions' in review panels, and the phenomenon of 'talking up or down' of proposals.]

At the micro-level of ongoing practices, there is variation and learning, embedded in a dual repertoire (or repertoires, there may be more than one external orientation). Such practices are bounded, and the boundary can be more or less strict. Thus, it is always possible to identify social worlds, even if fuzzy. Interlocking social worlds is how the meso-level is constituted, and enables and constrains micro-level practices.

Actors, while coming from a specific social world, may identify with an actual or emerging social world at the meso-level (perhaps macro-level). Example: Fred van

---

17 The recurrent reference to social worlds does not imply that the processes of exchange and trust that appear central to social worlds (as the locus of a community of practice) are the only processes that occur and are relevant. There are also strategic interactions, up to their stabilization in strategic games. And heuristics to be productive, like the "grammar" embedded in a regime that shapes further technological development.

Roosmalen (Philips), Roger de Keerschmaeker (IMEC) and Paolo Guardini (Intel) at the 2006 INC2 meeting try to get interaction going, without referring back to their organization and its interests. While they have a global argument that it is in the interest of their organizations that there will be such a shared world.

The other side of the meso-coin is how actors, spokespersons for their social worlds, take part in strategic games. Also when the game and its rules are still unclear, this happens, in the sense that there is strategic action which sees and uses the context as an opportunity structure.

Further example of Philips and General Electric in 1920s and 1930s: directors would meet and fight a battle of interests – the R-repertoire. And be accompanied by their engineers, who would visit their counterparts and learn about some of the directions in R&D and product development. This was actually expected, even if it should not be mentioned officially – the C-repertoire.

Such practices and how they function to create a working order may well have continued, and across all industries, to soften the hard edges of secrecy and intellectual property rights. Thus, *de facto* governance, as I will discuss it below (in Step 6). The recent interest in so-called open innovation implies that these governance questions are discussed explicitly: how to share intellectual property productively, and create exchange relations without too much strategic action. In the end, productivity will depend on the emergence and stabilization of a productive combination of C- and R-repertoire.

[Macro-level as resulting from such sharing of contexts? Also socio-technical landscape!]
[Culture as a patchwork of repertoires (cf. also Ann Swidler) on macro-level, to be mobilized by actors as a resource. Membership (i.e. social world) is still at issue, even if fuzzy, because actors can be strongly or weakly included. Use my 'split second' column as an example?]

Step 6

The preceding analysis can be used instrumentally, to be more effective in achieving one's goals, for example, in steering scientific and technological developments. Such goals may have a collective or public-interest component, for example when there is a diagnosis of an improductive lock-in which should be overcome, somehow. More generally, one can inquire into the quality of the interactions and emerging patterns.

The 'good firms' aligning themselves against the 'cowboy firms' are actually a cartel which attempts to exclude others so as to pursue the interest of its members. This may well serve a desirable collective goal, as in the earlier examples of safety and environment. But it is a cartel, and this is clear when big firms press for stricter environmental regulation, knowing that it will be difficult for smaller firms to comply. Regulation will then do away with their competitors for them.

In another version of the cartel, firms attempt to close their ranks to not give in to outside pressure. A clear example is how tobacco firms, in the 1970s, agreed not to use health aspects in their advertisements. Their joint interest was to present smoking as an acceptable habit. But then one company broke ranks, and advertised a "light" cigarette (light on tar and nicotine). After the first wave of indignant reactions, the other companies followed suit. We now live in a world in which "light" cigarettes are available. Cowboy firms are a danger to the order of the present social worlds. But sometimes this order should be undermined, and cowboys who do this, even if following their own interest, are then to be welcomed.

Normative debates, as about smoking and health and the responsibility of tobacco companies, have no easy solution. The outcome depends just as much on strategic actions and interactions of the various actors, as in the example I discussed above. Such contextual dynamics are important in all cases, and this is linked to the dual C-and R-repertoires. I will take euthanasia as a concrete case/example.

Originally, euthanasia was forbidden, but it occurred anyway, and was not reported. And when it was reported, the police officer wanted to find out if the medical doctor (often, a general practitioner) was a "good" medical doctor, and so could be excused. There was an enforcement and compliance world in place which (presumably) allowed acceptable practices.

The public discussion about liberalization of euthanasia was important, but it also induced reluctance of general practitioners, and medical doctors in general, to report on euthanasia, and actually perform euthanasia, even if it would help the patient and his/her family. The net effect is that there is less euthanasia. The social world of the medical practitioners is broken open by the liberalization discussion, third parties come in, professional nurses have an ambivalent role. The net effect is that medical doctors play it safe, and do less euthanasia. Which is the opposite of what was intended by the liberalization debate.

Recent debates on euthanasia. Data on reporting.

Flurry of news items end of April 2006, when the regional review committees published their annual report, and the undersecretary announced she wanted anonymized publication (on the Internet) of the cases and how these were judged, so as to make everything transparent, and show how careful/conscientious everything is handled.

The number of euthanasia reports (by medical doctors to the review committees) increases over the last years: 1815, 1886, 1933 in 2003, 2004 and 2005, respectively. One interpretation is increased willingness to report—because of assurance of careful handling. Another interpretation is increase in the actual number of cases of euthanasia, with the same willingness to report.

The regional review committees were introduced in 1998, in the hope medical doctors would report their euthanasias more often. The estimate then was that 50% was reported.]

This occurred at the time that a proposal for a law making euthanasia not a crime if done by a medical doctor who would follow certain requirements of scrupulousness [zorgvuldigheidseisen]. Such a law was eventually enacted in 2001.

At that time (April 2001), Minister Borst created a flurry of debate and concern when she said she thought that the "pill of Drion", a suicide pill, should be made available under well-circumscribed conditions. Interestingly, Drion's original proposal in 1991 was for a two-stage suicide means (a combination of two drugs each of which was not deadly by itself), not for a pill. But very soon, people spoke of the "pill of Drion", and that became the common reference. Again in Minister Borst's remark and the ensuing debate, even while Drion was interviewed and said he did not propose a pill (Volkskrant, 17 April 2001), and NRCHandelsblad (19 April 2001) added a box to one of its articles referring to Drion's original proposal and discussing the possibilities of it actually being produced.

The "pill of Drion" is a point representation (label) of a black box, and exerts force as a point representation. Cf. earlier on Chernobyl.

Back to 2006: Making the judgment of the cases by the euthanasia review committees available to all will improve understanding what zorgvuldigheidseisen mean in practice (says chairperson of the review committees). Professor Van der Wal (sociale geneeskunde) and evaluator of the euthanasia law sees this as a way to actively disseminate casuistics to the profession. (NRCHbl 27 April 2006)

A first-round conclusion is that strict enforcement of official rules is often improductive.

In other words, there are grey zones and these should remain so that situated judgement and action can occur. There are risks, of course, of limited and/or biased judgements, and undesrable shifts in the pattern of action – the "slippery slope" argument in the euthanasia debate. I discussed this issue already already (in Step 3) in terms of C- and R-repertoires.

Step 7

I am willing to argue that grey zones should be recognized as valuable. They constitute locations (zones) where repair work can take place to ensure that things go well on location. Such repair work includes adapting of (or sometimes forgetting about) general and thus never simply applicable rules. That our societies work, and continue to work, depends on such often invisible repair work rather than on the right organization and regulation.[18] Even while these have a function as well.

A second-round conclusion is that not all grey zones should be cherished. Grey zones are also locations for shady deals, which may be productive for those immediately involved, but not for wider society. Shady deals should be exposed (cf. also whistle blowing), and that can be done by reference to existing laws and justice, to R-repertoires more generally. But no blind quest for transparency, putting pressure on practices which will be counterproductive. Nor blind crime-hunting when whatever rules we have at the moment are trespassed.[19]

A working combination of C- and R-repertoire is productive, but must be checked for overall productivity. Just like narrow reflective equilibrium (Rawls, Daniels) must be checked by working towards a wide reflective equilibrium (Thagaard).

One could say that the picture I have drawn of how our society orders itself addresses its *de facto* governance (including governance through artifacts and infrastructures, not discussed in this text[20]). It is based on, and takes cognizance of, the granularity of the ordering of our world, including grey zones, interstices, new spaces, novelties and their precarious survival. Quality control (including legitimity) of such *de facto* governance must be dynamic quality control because the criteria and rules are not, *casu quo* do not remain, given.

"We've come a long way since Chernobyl." For philosophers, steps 6 and 7 will be the most interesting, after they have followed me on what one could call (and I have

---

18 I have argued this point using examples like the 2000 fireworks explosion in Enschede (in essays in Dutch). It is also visible in my analysis of the 'danger culture'of indutrial society (published in 1991).

19 These strong claims can be defended in detail. The tendencies described in the claims have to do with the dominance (and temptation) of narratives of praise and blame in our societies. These will also create point representations of "the" cause (=criminal or benefactor) – up to point representations of technologies (or technology in general) as the cause and/or remedy of our ills, and of chemicals as damaging health and environment, so regulating *them* will solve our problems.

20 But present in the back of my mind. See my analysis of technology as material narratives and integral to the constitution of our societies, in Chapter 7 of this book.

called) a sociological detour.[21] But it is not a detour, it is the main way. There is no clear destination (destiny?), however, and we make the way by walking.[22] But we have to be reflexive (philosophical?) about it.

---

21   In the lecture of April 1987, I continued to draw out implications for teaching and research in philosophy of science, technology and society (WWTS),

22   *'Caminante no hay camino, se hace camino al andar'* (Traveler, there is no path, the path is made by walking); Antonio Machado, Chant XXIX Proverbios y cantarès, Campos de Castilla, 1917. (Thanks go to Pierre Delvenne who checked this quote out.)

Printed in the United States
By Bookmasters